# History, Philosophy and Theory of the Life Sciences

Volume 13

**Editors**
Charles T. Wolfe, Ghent University, Belgium
Philippe Huneman, IHPST (CNRS/Université Paris I Panthéon-Sorbonne), France
Thomas A.C. Reydon, Leibniz Universität Hannover, Germany

**Editorial Board**
Marshall Abrams (University of Alabama at Birmingham)
Andre Ariew (Missouri)
Minus van Baalen (UPMC, Paris)
Domenico Bertoloni Meli (Indiana)
Richard Burian (Virginia Tech)
Pietro Corsi (EHESS, Paris)
François Duchesneau (Université de Montréal)
John Dupré (Exeter)
Paul Farber (Oregon State)
Lisa Gannett (Saint Mary's University, Halifax)
Andy Gardner (Oxford)
Paul Griffiths (Sydney)
Jean Gayon (IHPST, Paris)
Guido Giglioni (Warburg Institute, London)
Thomas Heams (INRA, AgroParisTech, Paris)
James Lennox (Pittsburgh)
Annick Lesne (CNRS, UPMC, Paris)
Tim Lewens (Cambridge)
Edouard Machery (Pittsburgh)
Alexandre Métraux (Archives Poincaré, Nancy)
Hans Metz (Leiden)
Roberta Millstein (Davis)
Staffan Müller-Wille (Exeter)
Dominic Murphy (Sydney)
François Munoz (Université Montpellier 2)
Stuart Newman (New York Medical College)
Frederik Nijhout (Duke)
Samir Okasha (Bristol)
Susan Oyama (CUNY)
Kevin Padian (Berkeley)
David Queller (Washington University, St Louis)
Stéphane Schmitt (SPHERE, CNRS, Paris)
Phillip Sloan (Notre Dame)
Jacqueline Sullivan (Western University, London, ON)
Giuseppe Testa (IFOM-IEA, Milano)
J. Scott Turner (Syracuse)
Denis Walsh (Toronto)
Marcel Weber (Geneva)

More information about this series at http://www.springer.com/series/8916

Malte Christian Ebach

# Origins of Biogeography

The role of biological classification in early plant and animal geography

Springer

Malte Christian Ebach
Biological Earth & Evironmental Sciences
UNSW
Kensington, Australia

ISSN 2211-1948 ISSN 2211-1956 (electronic)
History, Philosophy and Theory of the Life Sciences
ISBN 978-94-017-7991-3 ISBN 978-94-017-9999-7 (eBook)
DOI 10.1007/978-94-017-9999-7

Springer Dordrecht Heidelberg New York London
© Springer Science+Business Media Dordrecht 2015
Softcover reprint of the hardcover 1st edition 2015
This work is subject to copyright. All rights are reserved by the Publisher, whether the whole or part of the material is concerned, specifically the rights of translation, reprinting, reuse of illustrations, recitation, broadcasting, reproduction on microfilms or in any other physical way, and transmission or information storage and retrieval, electronic adaptation, computer software, or by similar or dissimilar methodology now known or hereafter developed.
The use of general descriptive names, registered names, trademarks, service marks, etc. in this publication does not imply, even in the absence of a specific statement, that such names are exempt from the relevant protective laws and regulations and therefore free for general use.
The publisher, the authors and the editors are safe to assume that the advice and information in this book are believed to be true and accurate at the date of publication. Neither the publisher nor the authors or the editors give a warranty, express or implied, with respect to the material contained herein or for any errors or omissions that may have been made.

Printed on acid-free paper

Springer Science+Business Media B.V. Dordrecht is part of Springer Science+Business Media (www.springer.com)

*Drawing lines on a map just to show I'm there*
*– Bear Cage/G.m.b.H (The Stranglers, 1980)*

# Prologue

*Biogeography means different things to different people depending, of course, upon their outlook and upon their biases (Ball 1975: 407)*

There is no scientific field more contentious, fragmented and misunderstood as biogeography. The confusion as to what biogeography is and how one defines it has presented both scientists and historians of science with a convoluted history. Do we define biogeography by ideas or by what biogeographers actually do? Many histories of biogeography (plant and animal geography), written by both historians of science and scientists, are about ideas, mostly about the development of evolutionary theories. Other histories concentrate on the results of biogeographical practice, such as theories derived from studying disjunct distributions. Few histories, however, focus on what biogeographers actually do and fewer still are aimed at scientists.

In this book I present a history of scientific practice, a *science historiography*, namely the history of methods used by plant and animal geographer between the 1770s and the 1890s. I have done this by rereading the historical scientific literature, rather than engaging in the literature written by historians of science. The point of a "science historiography" is to present the twenty-first-century biogeographers with a history they can understand and relate to in terms of how they practise their science. I am a scientist and a practitioner of biogeography and systematics, and I have written this book *primarily* for the science community. Historians of science may use this book as a secondary resource and as an opportunity to study the viewpoint of a twenty-first-century scientist.

The term *biogeography* was coined late in the nineteenth century separately in German and English, at a time when plant and animal geography were both firmly established but practised separately. The calls to unify plant and animal geography within a "biogeography" were made at a time when early forms of ecology (vegetation geography) and taxonomy were battling for a natural classification. This book looks at how the practice of classification has shaped early plant and animal geography before the 1890s.

This book will also examine how scientific specialisation has favoured certain forms of practice over others. For instance, the botanical geography of Augustin Pyramus de Candolle was in part admonished by certain plant geographers due to its apparent use of "artificial" regions, rather than "natural" vegetable formations: the former required training in taxonomy, geography and Earth history, while the latter was ahistorical, based on measurement, and therefore easier to do. Classification, a form of academic specialisation, influenced the practice of regionalisation, such as the mapping of taxonomic groups of plants and vegetation types (e.g. woodlands, savannahs).

*Origins of Biogeography* aims to explain the multiple origins of the nineteenth-century plant and animal geography (later called biogeography), to show that it was influenced by methods and practices from other fields and to demonstrate that the plant and animal geography practised in 1800 is vastly different from that in 1900 and 2000. I will also show that scientific practice and specialisation, rather than scientific ideas or theories, drove plant and geography between 1777 and 1900. A history of scientific practice may help a practicing twenty-first-century biogeographer understand the relevance and impact of classification on the history of their field. Focusing on the practice of classification, rather than on evolutionary or developmental theories or syntheses, presents a different history of plant and animal geography altogether.

Another aspect to this book is to challenge our current understanding of how biogeography was practised by engaging scientists with a history that they may not otherwise read. Recent histories of biogeography, written by scientists, are steeped in the search for the founders of ideas or acts as a tool, one that Australian historian and philosopher of science John S. Wilkins believes:

> ... has only an instrumental value in science. If it can be used to advance particular scientific claims then it is valuable; elsewise it is simply vapid, irrelevant or a waste of time and resources, according to scientists. (Wilkins 2014, p. 281)

I will attempt to debunk several of these histories, namely that plant and animal geographies have clear-cut origins:

> One can begin with Alexander von Humboldt (1769–1859), who is often recognized as the father of plant biogeography [...] [Wallace] was a key founder of zoogeography [...] Over the next 100 years the concepts of evolution and the role of dispersal in geographic patterns were expanded, but there was little real change in methodology. (Wen et al. 2013: 913)

Moreover, I challenge the idea that evolutionary theory was the expected outcome of the nineteenth-century plant and animal geography:

> From the beginnings of evolutionary thought in the early to mid-nineteenth century, geographic patterns figured prominently both as challenges to evolutionary explanation and by providing support for descent with modification. (Donoghue 2013, p. 76)

The practice of the nineteenth-century plant and animal geography was more an attempt to understand the world using taxonomic or vegetation classification, rather than the more popular stance of naturalists striving for a unified synthesis.

This volume is written primarily for scientists. The length is short, and the message concise. Each chapter may be read individually, every citation is written out in full and there are plenty of helpful footnotes, a reference section and an index. I have also included a glossary of historical figures, allowing for a comprehensive discussion without the distraction of introductions.

The text is divided into six parts that cover the late eighteenth and much of the nineteenth century. Chapter 1 introduces biogeography, an overview of the histories written by scientists and historians and the role of specialisation in the history of biogeography. Chapter 2 focuses on the role of distributional theories on the practice of animal and plant biogeography. Chapter 3 follows the development of early plant geography and the interactions between three of its practitioners, Augustin Pyramus de Candolle, Alexander von Humboldt and Friedrich Stromeyer, and how this had shaped the classification of plant geography. Chapter 4 follows the split between taxonomic and vegetation plant geographers, the rise of ecological animal geography and the rivalry between early ecologists and evolutionists over the role of natural classification. Chapters 5 and 6 will focus on regionalisation, the application of an area classification plant and animal biogeography. The end section of the book contains a glossary of historical figures mention in this book (Biosketches) – a handy guide for scientists wishing to know more about past practitioners featured in this book.

# References

Ball, I. R. (1975). Nature and formulation of biogeographical hypotheses. *Systematic Zoology, 24,* 407–430.
Donoghue, M. J. (2013). Historical biogeography. In J. B. Losos (Ed.), *The Princeton guide to evolution* (pp. 75–81). Princeton: Princeton University Press.
Wen, J., Ree, R. H., Ickert-Bond, S. M., Nie, Z., & Funk, V. (2013). Biogeography: Where do we go from here? *Taxon, 62,* 912–927.
Wilkins, J. S. (2014). Book review, do species exist? Principles of taxonomic classification. *Systematic Biology, 63,* 280–281.

# Acknowledgements

I would like to thank the following for reading through various drafts: John S. Wilkins, David M. Williams, Melinda L. Tursky, David Oldroyd and Tegan A. Vanderlaan. I also thank René Zaragüeta i Bagils and an anonymous reviewer for their invaluable comments, Mark Garland of *Scientific Latin Translations* for translating the introduction to Stromeyer's *Specimen*, Günther Beer for providing a high-quality image of Friedrich Stromeyer, Lynne R. Parenti for supplying high-quality images, Alan Kwan for photography, Gareth Nelson for discussions on the history of biogeography and for translating Humboldt from French into English and to the former students of GEOS3071 at UNSW Australia and BIZ5729 at the Universidade de São Paulo, Brazil, for listening to the history told in this book unfold in their lectures. I also thank Giesela R. Dohrmann and Rainer Dittmar for their help with German translation; Gareth Nelson and Brendan Doyle for French translation; and Felix, Vesper and Purley for their raspberries, meows and purrs. I am very grateful to my wife Melinda L. Tursky for her support.

# Contents

| | | |
|---|---|---|
| **1** | **A History of Biogeography for the Twenty-First Century Biogeographer**............................................................... | 1 |
| | Why We Need Another History of Biogeography .......................... | 1 |
| | A History of Histories...................................................... | 2 |
| | What Is Biogeography?.................................................... | 3 |
| | A Critique of Biogeographical Histories ................................... | 5 |
| | A History of Practicing Plant and Animal Geography ..................... | 15 |
| |     A Note on Colonialism ................................................. | 16 |
| |     Where Does It All Start? ................................................ | 17 |
| | References ................................................................. | 17 |
| **2** | **Origins, Race & Distribution** ............................................... | 21 |
| | Buffon's Law in Eighteenth Century Natural Classification ............... | 21 |
| | A Note on the Differing Themes of Distribution and the Roles of Naturalists ................................................................ | 26 |
| | Zimmermann's Legacy: From the *Specimen Zoologicae Geographicae* to Nineteenth Century Animal Geography ................ | 28 |
| | Distributional Laws ........................................................ | 36 |
| |     Climate as a Law of Distribution ....................................... | 38 |
| |     Vegetation and Faunas as a Law of Distribution ....................... | 39 |
| |     Regionalisation and the Law of Distributions .......................... | 40 |
| | References ................................................................. | 42 |
| **3** | **Humboldt, Stromeyer and Candolle** ...................................... | 47 |
| | Friedrich Stromeyer ....................................................... | 48 |
| | The Scope of Stromeyer's *Specimen* ...................................... | 49 |
| | Candolle's Grudge ......................................................... | 54 |
| | The Monopoly of the Ages: The Rise of Humboldt and the Fall of Stromeyer ............................................................... | 58 |
| |     Humboldt and de Candolle's Methods Compared ...................... | 64 |
| | The Geographies of Kant and Humboldt................................... | 67 |
| | References ................................................................. | 74 |

| | | |
|---|---|---|
| **4** | **Classification Divided**......................................................... | 79 |
| | Natural Classification and the "Two Courses" in Plant Geography......... | 80 |
| |     Literature-Based and Field-Based Research ............................ | 81 |
| |     Humboldt's Legacy and the Classification of Vegetation................. | 89 |
| |     Artificial and Natural Classification: A.P. Candolle and His | |
| |     Geography of Plants....................................................... | 95 |
| | Toward a Unification ......................................................... | 97 |
| |     Specialisation, Unity and the Proliferation of Terms .................... | 100 |
| | References ................................................................... | 104 |
| **5** | **Plant and Animal Geography in Practise: Maps, Regions** | |
| | **and Regionalisation**......................................................... | 107 |
| | Mapping Distribution (1777–1800) ........................................... | 108 |
| | Mapping Natural Areas (1805–1858) ......................................... | 113 |
| | A Multitude of Regionalisations (1858–1899)............................... | 125 |
| |     Plant Regionalisation ..................................................... | 125 |
| |     Animal Regionalisation .................................................. | 126 |
| | References ................................................................... | 136 |
| **6** | **The Legacy of Nineteenth Century Plant and Animal Geography** ..... | 143 |
| | The Multidisciplinary Nature of Biogeography............................. | 152 |
| | References ................................................................... | 154 |

**Epilogue** ............................................................................... 157

**Biosketches**.......................................................................... 159

**Appendix. Translation of the Introduction to "Commentatio Inauguralis Sistens Historiae Vegetablium Geographiae Specimen" by Friedrich Stromeyer (1800) (Translation by Mark Garland)** ................................................................... 165
    Introduction................................................................... 165
    References ................................................................... 172

# Chapter 1
# A History of Biogeography for the Twenty-First Century Biogeographer

## Why We Need Another History of Biogeography

Most histories written for scientists are aimed at identifying a founder, usually a patriarchal figure from whom all knowledge originates. While this may serve some practitioners of science to unify a field, it is in the most part a political exercise. Twenty-first century biogeography has multiple origins, most of which are in the twentieth century. Few, if any methodologies, theories and implementations of twenty-first century biogeography go back to the nineteenth century, let alone the eighteenth. What is more, twenty-first century biogeography has many different practitioners that hail from different backgrounds, very much like the practitioners in the eighteenth and nineteenth centuries. The calls for unity in biogeography in twenty-first century are remarkably similar to those in the late nineteenth century. Take any given number of practitioners from different backgrounds (e.g., taxonomy, geography) and allow them to pursue questions about organismal distribution, you will invariably end up with a multidisciplinary field regardless in which century you practise. The aim of this book is to show that eighteenth and nineteenth century plant and animal geography is a multidisciplinary profession and in as much conflict as twentieth and twenty-first century biogeography. The problems being encountered in biogeography today (e.g., calls for unification) are the same as those in the past. *Origins of Biogeography* is a history for twenty-first century biogeographers that detail the confusion of geographical and taxonomic laws (Chap. 2), the conflict between practitioners (Chap. 3), the divergence of classifications (Chap. 4), and the way we implement our plant and animal geographies (Chaps. 5 and 6) in the eighteenth and nineteenth centuries. By the end of the nineteenth century the basic divisions in twentieth century biogeography are already apparent. The twentieth century has its own unique history, which this book will not cover. Few biogeographers see twentieth century origins in their field.

## A History of Histories

Many recent histories of biogeography, as found in the introductory chapters of scientific texts or as articles in journals, tell of a glamorous beginning, usually in the works of a famous naturalist like Alexander von Humboldt, Charles Darwin or Alfred Russel Wallace. The origin of biogeography, we are told, was a result of "founders" and "fathers":

> Alexander von Humboldt, the father of biogeography, opened the way for ecological ideas after an expedition to South America ... (Humphries and Walker 2013, p. 5).

> [Darwin] framed most of the important questions that still define our understanding of evolution, from natural selection to sexual selection, and founded the main principles of the sciences of biogeography and ecology (Padian 2008, p. 634).

> The father of biogeography was Alfred Russel Wallace ... (Ellis 2010, p. 77).

However, there are fallacies implicit in these statements. What may catch the attention of some biogeographers and historians of science is the modernity of the terms used, retrospectively attributed to designated founders and fathers. The term "biogeography" was not in use until the end of the nineteenth century, being coined independently in German and English in 1883 and 1892 respectively (and the same is true for "ecology", which was coined in German in 1866 and in English in 1876),[1] and therefore post-dates the cited naturalists work. Could we for instance assume that the ideas of Wallace and Humboldt are of a direct intellectual lineage that have trickled down the ages to our very own research program? These histories have the tendency to lead to Whiggish claims of direct descent of ideas, making naturalists like Humboldt "biogeographers" or "ecologists" before the terms were coined (see Mayhew 2001 and Withers 2006). One may argue that Humboldt did do a form of "ecology"; however, ecology like biogeography has changed considerably within the last 200 years, with many biogeographers today practicing[2] a science that would have been unrecognisable 100 years earlier. These changes in biogeography are

---

[1] The first person I know of to use the term *œcologist* in English and in print was John Scott Burdon-Sanderson (1893, p. 465), in reference to Haeckel's original 1866 definition in *Generelle Morphologie der Organismen* (Haeckel 1866, p. 236). The term *ecology* first appeared in English in a translation of the same work in 1876. *Biogeographie* was coined by German zoologist, Hermann Jordan (1883, p. 174, although it is misspelt in the article title as "biographie") and, *biogeography* was coined by American mammalogist Clinton Hart Merriam (1892, p. 8). Neither Jordan nor Merriam defined "biogeography", suggesting that the terms were in use before the 1880s. The term *biogeographer* was used for the first time in English and in print in 1898 to describe the position of Charles Henry Tyler Townsend, "Biogeographer and Systematic Entomologist" (Anonymous 1898, p. 301). Müller (1996, p. 79) however states that Ratzel (1888) first coined the term "biogeographie" in a letter to H. Eisig in Nepal dated 31.1.1888. Clearly this is not the case.

[2] The term *practice* refers to the aims of a scientist as he or she goes about doing their science. For example, a nineteenth century plant taxonomist may wish to find out whether a genus of plants is in fact a natural group or not. To do this he or she would collect specimens, describe them, add them into collections, allocate new names or revise existing names (nomenclature), compare specimens and their parts (comparative anatomy) and place them into classifications (taxonomy).

rapid and are influenced by various different and often unrelated fields. In order to understand what biogeography means to biogeographers, it is well worth exploring what biogeographers attempt to do and how they are educated and trained.

## What Is Biogeography?

Biogeography today is generally defined as the study of the distributions of plants and animals over time.[3] To be more precise, early twentieth century biogeography is an amalgamation of late nineteenth century plant and animal geography that the nineteenth century German geographer Friedrich Ratzel saw as a fragmented discipline. Ratzel urged plant and animal geography be drawn together by a common principle. After all, oceanography and climatology are unified fields, why not biogeography? "It is the duty of geography ..." said Ratzel "... to go ahead and summarise and create a biogeography that shares a single common principle, to study the distribution of life on Earth" (Ratzel 1891, p. xxiv).[4] But few adopted Raztel's proposed solution.

Many practising plant and animal geographers during the eighteenth and nineteenth century were taxonomists and geographers who asked separate questions about taxonomy and geography. The same is true for those practicing biogeography today. For example, a molecular biologist will ask questions about the geography of genealogies within a population (e.g., phylogeography), whereas a population biologist may ask questions about the rates of colonisation of organisms onto an island (e.g., island biogeography), and an evolutionary taxonomist may wish to know the geographical history of their genus of plants across the globe (e.g., historical biogeography, palaeobiogeography). These practitioners may share a common theme (organismal distribution), but they lack unifying principals and methodologies.

To explore this further, we need to consider something few histories deal with: the machinations of biogeographers. For instance, what do biogeographers do? Why do they do biogeography, where do they get their ideas from and what sort of tools do they use? The answers to these questions depend on one's scientific background and training, the organisms one works on, and within in which period one happens to live.

Questions about the aims, education and training of the naturalists within any given period make biogeography a much harder field to pin down historically. A reason may be that biogeography, unlike other related fields, is multidisciplinary,

---

[3] See Platnick and Nelson (1978, p. 10), Morrone (2009, p. 7), and Millington et al. (2011). This definition also includes palaeobiogeography.

[4] In fact Ratzel saw biogeography as a geographical science. Once a zoologists asks "Where do these animal live?" and "What climatic factors and soils influence their distribution" then the science becomes geographical (see Ratzel in Müller 1992, pp. 451–452 and footnote 52).

one that describes individuals from different scientific backgrounds, who wish to ask questions about geographical distribution of their organisms, their parts (e.g., DNA) or the history of the areas in which their organism live or lived. With no specific aim or goal, other than general questions about distribution, biogeography has appeared fragmented.[5] Since 1891 there have been calls for biogeography to be unified or integrated by other more popular and specialised academic professions like geography (Joyce 2009; Milligan et al. 2011), anthropology (Harcourt 2012) phylogenetics (Riddle 2005; Donoghue 2011) and ecology (Rickleffs 2008). However, biogeography itself is so diverse that no one of these professions alone could encompass all aspects of its theory and methodology. As a result, no viable solution has presented itself since Ratzel, and it is unlikely that biogeography will ever be unified by any one of these fields, methods or theories. Biogeography today, like the nineteenth century plant and animal geography before it, is driven by its practitioners, their backgrounds, aims and skills. In this sense biogeography will always remain a multidisciplinary field.[6]

Moreover, the multidisciplinary aspect, themes and principals of biogeography need to include a temporal dimension. As these aims and skills change over time, the meaning and implementation of plant and animal geography of 1800 is vastly different from that of 1900 and 2000. Modern definitions of biogeography (or plant and animal geography) do not encompass the aims and goals of eighteenth and nineteenth century practitioners. While many biogeographers today are interested in studying the history of the geographical distribution of plants and animals, eighteenth and nineteenth century plant and animal geographers were cataloguing the diversity of nature. Naturally there were speculations among late eighteenth century naturalists like Carl Linnaeus, Heinrich Friedrich Link, Carl Ludwig Willdenow and Eberhard August Wilhelm von Zimmermann as to the possible machinations to distribution (see Larson 1994). But speculation is all that they were. After all, naturalists like Willdenow were practitioners of taxonomy with

---

[5]Michael Paul Kinch, commenting on the work nineteenth century naturalists, sums this up nicely: "The theoretical work which was produced often came not from the pens of empiricists in pursuit of biogeographical data, but instead from naturalists from various disciplines who were interested in the evidence biogeographical theory might provide for issues being debated in their particular fields" (Kinch 1980, p. 91).

[6]As I will show later in this book, ecology did derive from the vegetation (or ecological) plant geography of Grisebach, something that has been noted by various ecologists (Colinvaux 1973; Tobey 1981, see also Nordstrom 1990). A unique take on the relationship between ecology and biogeography is made Swiss botanist Eduard August Rübel "Geobotany (plant ecology-plant geography) is the science of the relationship of plants to the environment, the earth. [...] Historically plant geography, plant ecology, and geobotany are synonymous and include all six branches [(i) autochorologic geobotany; (2) autecologic geobotany; (3) autogenetic geobotany (combining with the study of the flora); and (4) synchorologic geobotany or chorologic sociology; (5) gynecologic geobotany or ecologic sociology; (6) syngenetic geobotany or genetic sociology, study of succession (combining with the study of vegetation or plant sociology)]. Geobotany (GRISEBACH) always does this; the two other terms are ambiguous, because often used in narrower and wider senses" (Rübel 1927, pp. 430, 437–438, original emphasis).

a primary interest in cataloguing observations of organismal distribution (either through personal experience or published travelogues), rather than investigating their origins.

Scientists who practise biogeography today also speculate about the processes that drive distribution. However, they do so by utilising principles from established professions like evolutionary biology, physical geography, macro-ecology and phylogenetic systematics. They use sophisticated interpretations of biological evolution (e.g., sympatry, allopatry) and physiological adaptation as well as phylogenetic relationships and varying degrees of geographical and geological models (e.g., tectonics, geomorphology). Together these form a whole new field of biogeography, with ideas and skills that were absent in the 1800s and early 1900s when animal and plant geography were being established. What, then, did these early naturalists do?

Quite simply, they catalogued organisms in *Floras*, zoologies or as travelogues, that is, monographic treatments of the classification and distribution of organisms as well as their physiology. Based on these *Floras* and their distributions, many attempted to divide up the planet into climatic, geographical or endemic areas that bound these floras and faunas. For example, naturalists engaged in taxonomy were interested in the distribution of taxa (e.g., species, genera) and sought to find natural provinces. Others like Humboldt were more interested in types of vegetation, the physical, chemical and biological factors that defined an area. The combination of different aims led to a multidisciplinary plant and animal geography, one that people like von Humboldt and de Candolle attempted to classify in order to stabilise and unify the field. However, neither of these classifications is with us today, even though they lead to new terms such as historical plant geography, phytogeography and biogeography. Instead, new classifications of biogeography appear today,[7] many driven by newly emerging fields that wish to investigate geographical distributions from the perspectives of their own specialisation.

## A Critique of Biogeographical Histories

The history of biogeography has been written mostly from an evolutionary viewpoint, that is, as a history of ideas:

> A reflective reader cannot help being surprised at the ease with which younger naturalists – Illiger and Treviranus, for example – concentrated upon purely systematic elements in the work of their predecessors and ignored their historical ideas (Larson 1994, p. 131).

By historical ideas, Larson means the "ideas concerning the historical development of nature", which he follows up with a chapter on "The Mechanism of Formation":

---

[7]For example, between 1999 and 2010, at least five classifications of biogeography were proposed based on methodology (Spellerberg and Sawyer 1999; Crisci et al. 2003; Morrone 2009; Parenti and Ebach 2009; Lomolino et al. 2010).

As controversy accumulated year after year, the system of *evolutio* established itself as the exclusive, universal, and specific mode of generation for living beings, opposed to any and all mechanist and materialist explanations (Larson 1994, p. 132, original italics).

What Larson is referring to here is the role of early ideas in "evolution" or *evolutio* "the unrolling of parts already existing in compact form" (Larson 1994, p. 132). The reader might also wonder why Larson is so concerned about the influence eighteenth century practitioners had on the "purely systematic elements in the work of their predecessors"? After all, Johann Karl Wilhelm Illiger and Gottfried Reinhold Treviranus both knew far more about taxonomy and physiology for instance than they did about "historical ideas" (Larson 1986, p. 488, 1994, p. 131). As naturalists they observed the world and pondered about questions and ideas that may seem unexciting to a historian or scientist who is interested in the development of evolutionary thought during the late eighteenth and early nineteenth century. While the topic may seem relevant to a twentieth and twenty-first century audience, it does not define the practise of early nineteenth century naturalists. This is not to discredit Larson or his work. Larson's chapter on the "Distribution of Natural kinds" concentrates on origins, a topic that eighteenth century naturalists merely speculated upon.[8] The majority of naturalists like Illiger and Treviranus focused on creating catalogues of plant or animal distributions and the environments where they are found, taxonomic keys, anatomical nomenclature, physiological descriptions, reproduction and ontogenetic development. While a history of what naturalist thought is both interesting and valid, it does not necessarily help twenty-first century biogeographers understand how their field or specialisation emerged from a more general natural history.

The perspective a writer of history takes, and the questions they ask, may greatly influence the interpretation of historical events. For instance, Larson (1986) presented a historiography of eighteenth century "geographical history". In it he posits, for example that "In the *Histoire naturelle* zoogeographical problems occupy so central a position that it is possible to view Buffon as a founder of biogeography" (Larson 1986 pp. 447–448, original italics). Larson here refers to American ichthyologist and systematist Gareth Nelson (1978), without explanation as to why such a "selective reading of Buffon", as Larson terms it, was chosen as a source. Nelson's *From Candolle to Croizat: Comments on the History of*

---

[8]Michael Wallascheck has dedicated nine volumes to the history of ideas in his *Fragmente zur Geschichte und Theorie der Zoogeographie* (Wallaschek 2009–2013). The aim of Wallascheck's monumental history is to analyse the "literature on the development of concepts, methods and theories of zoogeography" (Wallaschek 2009, p. 4, my translation), although it contains a new biogeographical synthesis and a new classification of zoogeography. Wallaschek's *magnum opus*, however, only focuses on the middle European German literature (with the exception of German translations of Buffon, Darwin, Wallace and modern biogeographers), and ignores virtually all histories of biogeography, with the exception of those written in German (e.g., Hofsten 1916; Schmithüsen 1985; Feuerstein-Herz 2004). While a review of the German literature on zoogeography is admirable it does however leave many gaps in the overall history of plant and animal geography.

*Biogeography* written during a time when there was a literal war between the established biogeographers, such as American entomologist Phillip J. Darlington (as self-professed biogeographer) and a new generation of systematists who read and were inspired by Lars Brundin, Leon Croizat and Willi Hennig.[9] Unbeknownst to Larson perhaps, is the reshaping of the systematic landscape.[10] Palaeontologists like Simpson who had dominated evolutionary biology were challenged by a whole new generation of cladists (phylogenetic systematists) who effectively levelled the playing field. In cladistic theory and practise, ancestors were no longer the starting point for ghost lineages that climbed up stratigraphic columns. Rather, they were treated in the same manner as extant species within a classification. For population dynamics, an early predecessor to population genetics, cladistics eliminated the need for ancestral populations and complex genetic theories. Cladistics, in a single method, could determine the evolutionary pathways of homologs (the parts of organisms) and create hierarchical classifications without any grand synthesis. With the Modern Synthesis thus threatened, an eventual turf war broke out in the pages of *Systematic Zoology* (and later *Cladistics*), which was popularised (some say dramatised) in David Hull's *Science as a Process* (Hull 1988), too much praise and criticism (see Farris and Plantick 1989; Donoghue 1990).

Cladists like Nelson, who represented the field of taxonomy and systematics, were in a position to claim a founder for their new version of biogeography. Rather than acknowledge Darwin and Wallace, who had by now become the "fathers" of evolutionary biology and the Modern Synthesis, Nelson claimed that status for Buffon, which Larson is correct in believing was through "selective reading". But this selective reading was acknowledged by Nelson to come from the nineteenth century geologist Charles Lyell. It was Lyell who originally claimed Buffon as a founder of nineteenth century animal geography. Buffon may be read selectively as he did mention the geographical distributions of animals in reference to taxonomic practise (see Chap. 2). However, this is a highly critical reading of Buffon when we consider that his Law was a way to justify the disjunct distributions of the same species. In other words, Buffon's Law is a *taxonomic* rather than a geographical law.[11]

---

[9] I would also include the recent history of Llorente-Bousquets et al. (2000) as one written by cladists, which was heavily influenced by Nelson (1978) and Nelson and Plantick (1981). Also see Craw et al. (1999).

[10] In fact, by the late nineteenth century there was a rift in plant geography between those that worked with taxonomic classifications (taxonomic geography) and those that used vegetation (vegetation geography) as the primary emphasis of their research (see Hagen 1986 and Chap. 4). Hagen's taxonomic and ecological plant geography is based on the practise of plant geography during the nineteenth century, while Nelson's historical and ecological biogeography is based on de Candolle's theoretical division between habitats and regions.

[11] Hofsten also saw the limitations of Buffon's contribution: "Buffon placed no great importance on plant geography; his ideas about the independence of vegetation on climate were not new, as they had previously been discussed by Linnaeus, although the honour is often attributed to Buffon" (Hofsten 1916, p. 248, my translation). The original reads: "Für die Pflanzengeographie hatte er keine größere Bedeutung; seine Ideen über die Abhängigkeit der Gewächse vom Klima

During the late 1970s and early 1980s, cladists were desperate to split from the highly established Modern Synthesis, which included evolutionary taxonomists like Matthew and ecologists like MacArthur. Moreover, the most outspoken proponent of the Modern Synthesis, Ernst Mayr was explicit: "zoogeography has had a similar fate very much like taxonomy. It was flourishing during the descriptive period of biological sciences. Its prestige, however, declined rapidly" (Mayr 1944, p. 1). The Modern Synthesis had saved taxonomy through "geographic speciation" and the "isolation of populations", just like ecology, which had "reached a level of maturity at which it is beginning to affect profoundly zoogeographic methods and principles" (Mayr 1944, p. 1). Almost 50 years later the cladists were keen to shed the shackles of the Modern Synthesis and, the further that cladists could distance themselves from Mayr, Simpson and Darlington, the better (see Williams and Ebach 2008; Hull 1988). Nelson's history did this by targeting Candolle's 1820 division between *stations* and *habitations*. A station was "the special nature of the locality in which each species customarily grows" and a habitation "a general indication of the country wherein the plant is native" (Candolle, in Nelson 1978, p. 280). Neither term was common in the plant or animal geographical literature, making Candolle's terms useful, in part, in dividing up present day biogeography:

> ... the concepts of station and habitation are important in Candolle's view, for they define two different sciences, which persist into the modern era [...] No matter, the terms as used by Candolle, have modern counterparts: ecological and historical biogeography. Ecological biogeography is the study of stations; historical biogeography, the study of habitations (Nelson 1978, p. 280, footnote 31, 281).[12]

With Buffon as the father of biogeography and Candolle seemingly its first practitioner, Nelson had effectively written a new history of biogeography for the cladists who exclusively practised historical biogeography.[13] Nelson's history is based on the history of classification. After all, twenty-first century historical biogeographers are, like their eighteenth and nineteenth century counterparts, also practicing taxonomists and systematists and, classification lies at the heart of what they practise.

---

waren ja nicht neu, sondern früher von Linné ausgesprochen, obgleich die Ehre vielfach Buffon zugeschrieben wurde" (Hofsten 1916, p. 248).

[12] Nelson's division is used by historians of science: "The descriptive, ecological side of biogeography was only one aspect of the inquires into distribution which began to occupy British naturalists in the middle decades of the nineteenth century. The other side was temporal, or historical, biogeography, encompassing investigations of how and why the distribution of species might have changed through time" (Rehbock 1983, p. 150). The mistake that Philip F. Rehbock makes is by assuming a twentieth century classification (ecological versus historical biogeography), based on a recent (1980s) demarcation within biogeography, to classify nineteenth century geography.

[13] Historical biogeography was limited to those who used cladograms in their study. Presently the term has deviated from its original meaning including non-cladistic fields such as macro-ecology.

Larson's history is taken from a general evolutionary perspective.[14] Nelson's history is exclusively from that of the cladist. Alternatively, it is possible to posit questions that appear common to multiple perspectives. One of the first twentieth century histories of biogeography written by Nils von Hofsten, a Swedish zoologist and anatomist, was entirely novel as it asked a question: "what are the origins of the fragmentation of endemic areas responsible for discontinuous [herein *disjunct*] distributions?"[15] (Hofsten 1916, p. 1). What makes Hofsten's history unique is the assertion that disjunction was a problem that equally puzzled all practitioners of animal and plant geography through the millennia as far back as Hippocrates. However, asking a universal question has its problems: are all practitioners using the same definition of the same thing? For example, Buffon had a very different conception of species than that of Linnaeus. Is the questioned posed in an evolutionary context, where geographical isolation plays a central role? Are the organisms referred to the same thing (e.g., species)? Each of these factors alters the question. For example, a twenty-first century biogeographer would take disjunction as a given and ask which tectonic, physiological or climatic mechanisms are responsible. Moreover, they would have vastly different materials and methods to hand. These concerns are moot once we read Hofsten as a scientist's historiography, one that catalogues and lists every author and text that had investigated organismal distribution – a taxonomy of animal and plant geographers and their work. It is important to note that Hofsten's work is not a simple chronology, of people and their contributions, like some modern treatments (Reed 1942; Schmithüsen 1970, 1985; Papavero et al. 1997, 2013). Hofsten introduced each practitioner of plant and animal geography and how they approached problems in distribution (e.g., disjunction, dispersal etc.). Reading Hofsten, one can see his discontinuity problem expanding to include theories about distribution (primarily dispersal), continental drift, evolution, climate; in effect encompassing all problems and causal processes ever asked in biogeography. While this might be a naive way to approach a field, it does circumvent specialisation. Hofsten was not arguing from any one field hoping to unify it into plant and animal geography. Rather he was concerned with distribution wholesale, thereby avoiding any association with a particular scientific discipline. Unity, it seems, is in distribution and all its assorted problems and theories. Hofsten, like Harvard plant ecologist Hugh Raup, attempted a scientific history, one that showed how scientific ideas (usually causations) were used to solve problems throughout the history of animal and plant geography.

Raup (1942) sums up plant geography as a field with multiple histories that diverged in the late nineteenth century on the basis of the classification of plants (see Chap. 4). As an ecologist Raup was aware of the ways in which eighteenth and nineteenth century naturalists were able to make sense of plant distributions

---

[14] Kinch (1980) also provides a Modern Synthetic history of nineteenth century plant and animal geography.

[15] Original German: "Welches sind die Ursachen dieser Zersplitterung der Heimat, dieser diskontinuierlichen Verbreitung?"

based on the materials and methods that were available. Unlike Hofsten, Raup acknowledges that plant geography and the "arrangement of its vast body of knowledge [...] rests, first, upon a clear understanding of the kinds of plants" (Raup 1942, p. 322). The classification of kinds of plants, however, was what made plant geography problematic. Humboldt and his contemporaries, Stromeyer and de Candolle, used different classifications for plant kinds. Humboldt preferred plant forms (e.g., the tamarind form, the lianas), while Stromeyer and de Candolle used Linnaean taxonomy to distinguish between types of plants and their distributions. Raup's thesis is simple: the diversity of plant geography, from its foundation to today was in how plants are classified.[16] However, Russian botanist Evgenii Wulff approached the topic of plant geography entirely differently. Rather than look at how plants were classified, Wulff (1932, 1943) returned to an age-old practise of classifying plant geography.

Wulff's history is unusual in that it is heavily reliant on classification. (After all he was a taxonomist!) His own classification of geography, reinterpreted from older classifications of plant geography by Humboldt, Stromeyer and Schouw, is used *a posteriori* to explain the diversity of plant geographies, particularly between plant geography and the history of plants. Translated from the original Russian, Wulff's *An introduction to historical plant geography* (1943)[17] draws on Stromeyer's classification, namely, the "Geography of Vegetables, or Phytogeography; Geographical History of Vegetables and Applied Geographical History of Vegetables". Stromeyer defined Geographical History of Vegetables (plants) as:

> Whether Vegetables formerly occupied the earth in the same way as at the present day, or if that is truly the case, and what they may have undergone before they arrived at that station in which they are now found; what causes provide a place for these changes, and what things follow from this (Stromeyer 1800, pp. 14–15).

Compare this to Wulff's definition:

> Historical plant geography has as its aim the study of the distribution of species of plants now existing and, on the basis of their present and past areas, the elucidation of the origin and history of development of floras, which, in turn gives us a key understanding of the earth's history (Wulff 1943, p. 1).

Wulff ascribes his definition partly to Stromeyer, but chiefly to Schouw, which he then in turn ascribes to Alphonse de Candolle, Adolf Engler, August Grisebach and Oscar Drude.[18]

---

[16] However, this idea was not new. Erik Nordenskiöld (1936, pp. 560–561) noted that there were two plant geographies by the late eighteenth century. See Chap. 4 and Nicolson (1987, p. 167).

[17] Hugh Raup wrote the forward to the 1943 translated version in which he directs the reader to Reed's *A short history of the plant sciences* (Reed 1942) and Greene's *Landmarks of botanical history* (Greene 1909). Both works are discussed below.

[18] Wulff lists the branches on plant geography into floristic, ecological and historical in a table, in which the terms of Stromeyer are spilt, most notably the Geography of vegetables (as floristic) and phytogeography (as ecological), which are given the same definition (see Luna-Vega 2008).

Apart from his attempts to re-classify plant geography as historical (rather than ecological), Wulff's history follows that of Hofsten, namely, chronicling the works of notable figures like Lyell, Darwin and Hooker as well as investigating the causes of plant distributions, such as dispersal, geographical isolation and speciation. However, it is to plant classification that we return in order to understand why both historians like Erik Nordenskiöld and, later Joel Hagen and Malcolm Nicholson, believe that plant geography had diverged in the early nineteenth century.

As I have shown above, both the historians and scientists acknowledge the split between the classification of vegetation and taxa. While scientists like Larson qualify the divergence through scientific methodology, it is the historians who have brought in an interesting philosophical argument, namely that of natural and artificial kinds. By introducing this Nicholson provides a historio-philosophical take on plant geography (in contrast to the taxonomic, ecological, cladistic and the historio-evolutionary views of Wulff, Raup, Nelson and Larson respectively). Nicholson, citing Cannon (1978), sees Humboldt's plant geography, and the botanists he influenced such as Schouw, as being a unique form of early nineteenth century enquiry that "is impossible to fit into any of our twentieth century disciplinary boundaries" (Nicolson 1987, p. 167). Unwittingly, Nicolson has effectively barred professionalism from claiming early nineteenth century founding texts (see Chap. 4). Severing the connection with early nineteenth century science leaves both twenty-first century scientists and historians of science in a quandary: what relevance do the works of, Willdenow, Humboldt, Stromeyer and Schouw have to modern history?[19]

Nicolson suggests that Kant's *Physische Geographie* introduces the idea of a natural system, one that is based on real phenomena rather than arbitrary groups. This, Nicolson argues, assumed "the existence of a functional inter-relation between all the individual phenomena of the earth's surface", most importantly, it led Kantian geographers to postulate "an underlying causal unity of Nature" (Nicolson 1987, p. 171). In other words, artificial classifications like those of Linnaeus were "out", and those of natural classifications of Jussieu and de Candolle were "in". But this was not so easy as it sounds.

Natural classifications are difficult to recognise, particularly for early nineteenth century taxonomists and geographers. Many taxonomic groups, species even, were considered to be artificial. The split between taxonomy and other natural units of organisms had started to form. Nicolson had cleverly cemented Nordenskiöld's claim[20] by using both philosophical and scientific arguments. The split in classifying organisms either as taxa (species, genera, families etc.) or as vegetation (savannah,

---

[19]Lynn K. Nyhart also questions the lack of an early twentieth century history of biogeography: "even less historical research has been undertaken on the history of biogeography in the early twentieth century than on ecology" (Nyhart 2009, p. 324). As I will show in Chap. 6, much of the biogeography early twentieth century *was* mostly ecology and is covered by historians of ecology.

[20]"In the field of plant geography, research has taken especially two courses, a systematical, which is ultimately based on Linnæus's observations and theories in connextion with the distribution of the plant species, and a morphological, which has its origin in Humboldt's theories on the

rain-forests etc.) had defined and separated nineteenth century plant geography into two distinct studies: the distribution of taxonomic groups and the distribution plant forms (vegetation). Unwittingly, however, Nicolson has found a similar division to that of Nelson along a completely different line of enquiry (i.e., Nelson used the size and causes with areas of distribution to differentiate between ecology and historical biogeography).[21]

Ecological histories, like that of Nicolson, have helped to define early nineteenth century plant geography. The most ambitious is that of American historian of science Frank N. Egerton (2012). Based on an astonishing 40 articles that appeared in the *Bulletin of the Ecological Society of America*, Egerton's *Roots of Ecology: From Antiquity to Haeckel* he explores Linnaeus' economy of nature, a popular topic in modern ecological studies. Unlike most such histories, Egerton explores the role of presentism (in the form of Whiggish histories) in tracing connections between antiquity and modern times. By bringing up the subject of Whiggish history, Egerton makes an interesting remark, "[m]any books written by professional historians of biology [...] are analytical within a non-Whiggish context, which is interesting to other historians of biology, but not often to ecologists, and such books do not stay in print" (Egerton 2012, xii). Egerton seems to suggest that scientists are interested in making connections between modern ideas and those of the "Greats" of the past. A good story does engage the scientist as well as the historian as Egerton suggests, but it does not justify making dubious connections.

The connection Egerton draws between Linnaeus' economy of nature with a modern interpretation is typical of most history of ideas that stem back beyond modern times. Generally, an idea that is very popular today, be it evolution or the economy of nature, is traced back to the origin of the covering term, in this case Linnaeus. The economy of nature for example is presently an exclusively ecological term for an idea with an established theory and supporting literature, for example, the recent student textbook *The Economy of Nature,* now in its sixth edition. What Linnaeus, a taxonomist, thought about the economy of nature has little bearing on its present meaning or usage, which has its origins in the more modern day literature?[22]

Egerton, like Hofsten, wants to see science as a continuum of ideas that can be traced back to antiquity. While this does produce a well-researched and written history, compelling to both scientists and historians alike, the connections may be tenuous. Present day scientists want to know about their origins, founders and founding ideas, and they want them traced, preferably to historical giants like

---

morphological association of different vegetable types with different countries and forms of landscape" (Nordenskiöld 1936, p. 560). These two courses will be explored further in Chap. 4.

[21]For a different usage of the terms ecological and historical biogeography, see Nyhart (2009), who uses the terms in a descriptive rather than in the formal manner (*sensu* Nelson 1978). The different may be due to a misreading of the historical chapter in Cox and Moore (2005).

[22]Rickleffs (2008) cites Cronon's *Uncommon Ground* in his introduction, suggesting that modern ecological studies have their origins in more modern ideas like environmentalism and statistical analysis.

Haeckel, Humboldt and Theophrastus. In 2008 American palaeontologist Kevin Padian claimed that "[Darwin] framed most of the important questions that still define our understanding of evolution, from natural selection to sexual selection, and founded the main principles of the sciences of biogeography and ecology" (Padian 2008, p. 634). Connecting what scientists do to prominent historical figures like Darwin, provide them perhaps with a sense of worth and significance. More than likely, present day scientists will find that the origins of their field and ideas lie elsewhere, in the more recent literature or in the ideas of contemporaries from a far less glamorous background.

Histories written by scientists tend to find tenuous connections between modern practise and historical figures (e.g., Cox and Moore 2005, 2010; Lomolino et al. 2010; Blumler et al. 2011), while others concentrate on who did what within large biographical chronologies, (e.g., Reed 1942; Papavero et al. 1997, 2013 and Schmithüsen 1970, 1985). The biographical history of de Candolle and Humboldt written by historians of science like Jean-Marc Drouin (in Acot 1998; Drouin and Huet 2002; Drouin 2004), or the history of cartography and biogeography by American historian of science Jane Camerini (1993, see Chap. 5), for example, tell of a different history, one in which science is connected to society, issues and events of the day.[23] They actually explain what early naturalists were doing and what ideas were important to them, rather than their connections to modern practitioners.

Conversely, one may argue that these tenuous connections are driven by particular fields or in "tradition building". American geographer Richard Hartshorne presents an excellent case study of specialisation in the sciences. Geographers, he laments, "are wont to boast of their subject as a very old one, extending, even as an organized science, far back to antiquity. But often when geographers in this country discuss the nature of their subject, whether in symposia or in published articles, one has the impression that geography was founded by a group of American scholars at the beginning of the twentieth century" (Hartshorne 1939, p. 198).

The re-invention of geography in Hartshorne's view was due to the assumption that "geography is some kind of knowledge concerned with the earth", but as geographers they can "endeavor to discover exactly what kind of knowledge it is" (Hartshorne 1939, p. 205). Geography was becoming established by the early twentieth century, with professors of geography to take up the reigns. However, Hartshorne was philosophical on the matter,

> What ever geography is, its venerable if not honored age would nullify the most enthusiastic efforts of any students to make it over into something entirely different. Geography is not an infant subject, born out of the womb of American geology a few decades ago, which each new generation of American students may change around at will (Hartshorne 1939, p. 205).

---

[23] Stephen Jackson (2009) presents an interesting and in depth history of Humboldt and Bonpland's *Essay on the Geography of Plants*. The text sets Humboldt's work within a nineteenth century context, with an excellent description of his work, methods, materials and aims. Also Nils Robert Güttler (2011) presents a good history of maps after the Golden Age of first generation plant geographers.

I disagree with Hartshorne here on a purely methodological matter. Most, if not all ideas, in science have modern origins that are based on recent innovations. Biogeography today uses molecular data and phylogenetic methodologies that have come as a result of recent innovations and breakthroughs of the 1960s and 1970s. The interpretation of the results, however, may have older origins, that date back to late nineteenth century ideas.

Specialisation is as important to scientists as it is to historians of science. After all, from where do historians of science get their ideas? Ideas that are popular in science today are what attract historians of science interested in the history of ideas and the naturalists that played leading roles. Histories such as these have led to a tenuous connection between historical figures, like Theophrastus and Humboldt (see Greene 1909), and the works of historical figures and current topics in science (e.g., Larson 1994). One history that is worth mentioning is that of the *Secular Ark: Studies in the History of Biogeography* by English historian of science Janet Browne.

Browne's (1983) *Secular Ark* is more a history of ideas, namely evolution (in light of recent developments in evolutionary biogeography) rather than a history of biogeographical practise.[24] Her intention is to link the ideas of present day biogeography with that of past historical figures. Her thesis is that naturalists weighed down by the idea of Noah's Ark and biblical notions of biological origins, were cured by the development of Darwinian evolution. Browne explains that by the turn of the nineteenth century, the idea of natural kinds was still tainted by Biblical ideas of creation (without referring to Kant's teleology). Moreover she claims that nineteenth century naturalist were inclined to catalogue numerical surveys, accumulating "raw material for some long-awaited Newton of the biological sciences, a man who would discover great universal truths in their work" (Browne 1983, p. 80). Geologists and palaeontologists who were unmoved by regionalisation or topographical study "preferred to wait for the formulation of a set of rules about distribution that could be applied to the fossil record" (Browne 1983, p. 86). The whole theme of the book is to set up Darwin and Wallace and their ideas on evolution, as "the foundation stones on which all modern biogeographical endeavour is based" (Browne 1983, p. 165). Apparently, without Darwin or Wallace, "no history of the subject could be written" (Browne 1983, p. 165). Browne is correct in thinking that evolution was the logical result of plant and animal geography. However, as a history of ideas, it fails to address why some biogeographical practises have changed little

---

[24] And I use the term *biogeography* loosely here. In several places Browne, like Egerton (2012) refers to terms not used at the time (e.g., ecology, biogeography). Browne uses the term *phytogeography* to describe the 1778 work of Johann Reinhold Forster, even though Stromeyer coined it in 1800. Egerton uses the term *ecology* (coined but not defined in 1861) to describe early nineteenth century naturalists, as does Worster (1977, see Cittadino 1979, p. 45). The most presentist claim is possibly that of Einar du Rietz "[Linnaeus] was a pioneer not only in taxonomy and morphology but also in genetics, dispersal ecology and phytogeography" (du Rietz 1957, p. 161).

while ideas have changed drastically.[25] For example, recent developments in geospatial information systems as well as the new advances in computational power, have succeeded in quantifying Wallace's zoogeographical regions based on a large distributional database, yet their aims and results differ little to that of Wallace.[26] Compare this to the proliferation of ideas, such as lateral gene transfer, hybridism, ecological stranding, neotectonic events, that have recently been proposed to explain large zoogeographical distributions. A history of ideas may reveal little about why some methods like regionalisation have not greatly changed in 150 years.

If we were to instead look at how biogeographers do biogeography, then we may be able to, as Hartshorne put it, "discover what biogeographical knowledge is", in particular, how eighteenth and nineteenth century naturalists practised their plant and animal geography.

## A History of Practicing Plant and Animal Geography

I wish to present a new history: a history of doing plant and animal geography as seen by a scientist. The premise is that such a history will investigate the origins of ideas of the time, the knowledge possessed by each actor and how they attempted to address their aims and goals using the materials and methods available to them at the time.

I will also confine my history to the late eighteenth and much of the nineteenth centuries with my aim being to show that plant and animal geography were already established fields intellectually, and to show how this had influenced the ideas and methods of each practitioner. The result will be a practical history of plant and animal geography based on what was practised at the time. This practical history will also show that there are few connections between past naturalists of the eighteenth, nineteenth and twentieth centuries.

At this point a historian of science may ask: "what historical methods are employed in this study?" Rather than engage in the historical literature and write a historiography in the manner of a historian of science, I have written a *science historiography*, one that addresses the historical scientific literature. If historiography looks at the methods of historians and their written history, then a science historiography looks at the methods of scientists and their written history. Given this, I have not deeply engaged with the historical literature of historians of science

---

[25]Other history of ideas include Penny van Oosterzee's *Where Worlds Collide: The Wallace line* (Oosterzee 1997), Peter Bowler's *The Earth Encompassed: A History of the Environmental Sciences* (Bowler 1993), David Quammen's *The Song of the Dodo: Island Biogeography in an Age of Extinctions* (Quammen 1997) and more recently Dennis McCarthy's *Here be Dragons* (McCarthy 2009) and Alan de Querioz's *The Monkey's Voyage* (de Querioz 2014).

[26]Holt et al. (2013) have managed to map the known distributions of 21,037 species of vertebrates in order to create a quantifiable map of the zoological regions of the world.

nor employed any historical method. I refer the historian of science to the scientists who have done exactly this: Adolf Engler, Nils Hofsten, Erik Nordenskiöld, Evgenii Wulff, Nelson Papavero, Robin Craw, Gareth Nelson, David M. Williams, Olivier Rieppel and Jorge Llorente Bousquets. These scientists employed no standard historical method nor did they engage deeply with the literature written by historians of science, at least not wittingly.[27] My history follows in their tradition, a history of science through time that looks at the methods employed by plant and animal geographers in the late eighteenth and nineteenth centuries, namely a science historiography.

## A Note on Colonialism

Historians of science may note the absence of a discussion on European colonialism, institutions and societies and their effects on regionalisation and early plant and animal geography. A discussion on colonialism may reveal who collected specimens and measurements, where these specimens are held and which nations had a strong biogeographical tradition. Colonialism is a strong driving force in regionalisation. The expedition of Robert Brown on-board the *HMS Investigator*, for example, not only provided British institutions like Kew and the British Museum with a wealth of specimens, but it was also there to record the colonies mineral and biotic wealth (Flinders 1814; Ferrar 1984; see also Ebach 2012). Brown, one of the botanists in charge of collecting and recording part of Australia's biota, was the first to write about the links between Austral floras,

> The character of the New Zealand Flora, known to us chiefly from the materials collected by Sir Joseph Banks, is to a considerable degree peculiar; it has still however a certain affinity to those of the two great countries between which it is situated, and approaching rather to that of Terra Australis, than of South America (Brown 1814, p. 589).

Present day biogeographers may not be aware of the considerable impact European colonialism had on early plant and animal geography and the extensive historical literature. I refer them to Browne (1992, 1996) and Verran (2009) for detailed accounts on the effect of European colonialism on eighteenth and nineteenth century biogeography. While colonialism is important in facilitating the collections and measurements used by biogeographers, it has not guided its aims and goals. The nineteenth century pursuit of natural regions, for instance, is not influenced by colonialism. For example, the nineteenth century debate whether or not North America (Nearctic) and Europe and Eurasia (Palaearctic) belong in the same region (Holarctic) is based solely on the distribution of certain widespread

---

[27]Gareth Nelson claims no historical method was used in his often-cited 1978 article *From Candolle to Croizat: Comments on the history of biogeography*: "For me the problem was learning about an earlier history, which had been largely forgotten or ignored in the literature of my time [e.g., Mayr, Simpson, Darlington and so on]" (Nelson, 2014, personal communication).

mammals (see Chap. 5). Alfred Russel Wallace's insistence that the Nearctic and Palaearctic should be kept as separate regions has greatly influenced nineteenth, twentieth and twenty-first century zoogeographical regionalisation. Given this, I have not engaged in any discussion on the role of colonialism in eighteenth and nineteenth century biogeography.

## *Where Does It All Start?*

I will not start with the first mentioned use of plant or animal geography, nor with the first practitioners, like Aristotle's pupil Theophrastus, as is done in previous histories. Plant and animal geography or versions thereof, were done by a great many people through the ages. Rather I wish to look at the beginnings (or origins) of eighteenth and nineteenth century plant and animal geography, in the discussions of its practitioners. Starting at the "first ever" or "founding father" has its downsides, mostly that the first texts or founders rarely converse or debate their ideas publicly or in print. Instead, I will look at the interactions between people, their ideas and the environment that drives them. Most practitioners of plant and animal geography during the late eighteenth and nineteenth century were studying classification of some form. Therefore the debates I will discuss throughout this book will be about classification.

## References

Anonymous. (1898). University and education news. *Science, 192*, 301.
Blumler, M. A., Cole, A., Flenley, J., & Schickhoff, U. (2011). History of biogeography thought. In A. Millington, M. Blumler, & S. Udo (Eds.), *The SAGE handbook of biogeography* (pp. 23–42). Los Angeles: Sage Publications.
Bowler, P. J. (1993). *The earth encompassed: A history of the environmental sciences*. New York: W.W. Norton.
Brown, R. (1814). General remarks, geographical and systematical, on the botany of Terra Australis. In M. Flinders (Ed.), *A Voyage to Terra Australis; Undertaken for the purpose of completing the discovery of that vast country, and prosecuted in the years 1801, 1802 and 1803, in his Majesty's ship the Investigator, and subsequently in the armed vessel Porpoise and the schooner Cumberland* (pp. 533–591). London: G. & W. Nicol.
Browne, J. (1983). *The secular ark: Studies in the history of biogeography*. New Haven: Yale University Press.
Browne, J. (1992). A science of empire: British biogeography before Darwin. *Revue D'histoire Des Sciences, 45*, 453–475.
Browne, J. (1996). Biogeography and empire. In N. Jardine, A. Secord, & E. Spary (Eds.), *Cultures of national history* (pp. 305–321). Cambridge: Cambridge University Press.
Burdon-Sanderson, J. S. (1893). Inaugural address. *Nature, 48*, 464–472.
Camerini, J. R. (1993). Evolution, biogeography, and maps: An early history of Wallace's Line. *Isis, 84*, 700–727.

Cannon, S. F. (1978). *Science in culture: The early Victorian period*. New York: Science History Publications.
Cittadino, G. (1979). Nature's economy: The roots of ecology by Donald Worster. *Environmental Review, 3*, 45–47.
Colinvaux, P. A. (1973). *Introduction to ecology*. New York: Wiley.
Cox, C. B., & Moore, P. D. (2005). *Biogeography: An ecological and evolutionary approach* (7th ed.). Oxford: Blackwell.
Cox, C. B., & Moore, P. D. (2010). *Biogeography: An ecological and evolutionary approach* (8th ed.). Malden: Wiley-Blackwell.
Craw, R., Grehan, J. R., & Heads, M. J. (1999). *Panbiogeography, tracking the history of life*. Oxford: Oxford University Press.
Crisci, J. V., Katinas, L., & Posadas, P. (2003). *Historical biogeography: An introduction*. Cambridge, MA: Harvard University Press.
de Querioz, A. (2014). *The monkey's voyage: How improbable journeys shaped the history of life*. New York: Basic Books.
Donoghue, M. J. (1990). Sociology, selection, and success: A critique of David Hull's analysis of science and systematics. *Biology and Philosophy, 5*, 459–472.
Donoghue, M. J. (2011). Historical biogeography. In J. B. Losos (Ed.), *The Princeton guide to evolution* (pp. 75–81). Princeton: Princeton University Press.
Drouin, J.-M. (1998). Augustin-Pyramus de Candolle. In P. Acot (Ed.), *The European origins of scientific ecology, Volume 1* (pp. 359–422). Amsterdam: Overseas Publishers Association.
Drouin, J.-M. (2004). Introduction. In J.-D. Candaux & J.-M. Drouin (Eds.), *Augustin-Pyramus de Candolle – Mémories et souvenirs (1778–1841)* (pp. 1–35). Geneva: Georg.
Drouin, J.-M., & Huet, E. (2002). Biogéographie et systématique dans les Plantes équinoxiales d'Alexandre de Humboldt et Aimé Bonpland (1808–1809). *Bulletin d'histoire et d'épistémologie des sciences de la vie et d'épistemologie des sciences de la vie, 9*, 25–38.
Du Rietz, G. E. (1957). Linnaeus as a phytogeographer. *Vegetatio, 7*, 161–168.
Ebach, M. C. (2012). A history of biogeographical regionalisation in Australia. *Zootaxa, 3392*, 1–34.
Egerton, F. N. (2012). *Roots of ecology: Antiquity to Haeckel*. Berkeley: University of California Press.
Ellis, R. J. (2010). *How science works: Evolution: A student primer*. Dordrecht: Springer.
Farris, J. S., & Platnick, N. I. (1989). Lord of the flies: The systematist as study animal. *Cladistics, 5*, 295–310.
Ferrar, A. N. (1984). The graphical records of Matthew Flinder's voyage to Terra Australis. *Journal of Navigation, 37*, 94–103.
Feuerstein-Herz, P. (2004). Eberhard August Wilhelm von Zimmermann (1743–1815) und die Tiergeographie (pp. 1–389). Ph.D thesis, Technischen Universität Carolo-Wilhelmina zu Braunschweig.
Flinders, M. (1814). *A voyage to Terra Australis*. London: G. & W. Nicol.
Greene, E. L. (1909). *Landmarks of botanical history: A study of certain epochs in the development of the science of botany, Part I. Prior to 1562 A.D.* (Miscellaneous Collection, No. 1870). Washington, DC: Smithsonian Institution.
Güttler, N. R. (2011). Scaling the period eye: Oscar Drude and the cartographical practice of plant geography, 1870s–1910s. *Science in Context, 24*, 1–41.
Haeckel, E. (1866). *Generelle Morphologie der Organismen: Allgemeine Grundzüge der organischen Formen-Wissenschaft, mechanisch begründet durch die von C. Darwin reformierte Decendenz-Theorie.*. Berlin: G. Reimer.
Hagen, J. (1986). Ecologists and taxonomists: Divergent traditions in twentieth-century plant geography. *Journal of the History of Biology, 19*, 197–214.
Harcourt, A. H. (2012). *Human geography*. Berkeley: California University Press.
Hartshorne, R. (1939). The nature of geography: A critical survey of current thought in the light of the past. *Annals of the Association of American Geographers, 29*, 173–412.

# References

Holt, B. G., Lessard, J. P., Borregaard, M. K., Fritz, S. A., Araújo, M. B., Dimitrov, D., Fabre, P. H., Graham, C. H., Graves, G. R., Jønsson, K. A., Nogués-Bravo, D., Wang, Z., Whittaker, R. J., Fjeldså, R. J., & Rahbek, C. (2013). An update of Wallace's zoogeographic regions of the world. *Science, 339*, 74–78.

Hull, D. L. (1988). *Science as process: An evolutionary account of the social and conceptual development of science*. Chicago: University of Chicago Press.

Humphries, P., Humphries, P., & Walker, K. F. (2013). The ecology of Australian freshwater fishes: An introduction. In P. Humphries & K. F. Walker (Eds.), *Ecology of Australian freshwater fishes* (pp. 1–24). Collingwood: CSIRO Publishing.

Jackson, S. T. (2009). Introduction: Humboldt, ecology, and the Cosmos. In A. von Humboldt & A. Bonpland (Eds.), *Essay on the geography of plants* (edited by S. T. Jackson) (pp. 1–52). Chicago: University of Chicago Press.

Jordan, H. (1883). Zur Biogeographie der nördlich gemäßigten und arktischen Länder. *Biologisches Centralblatt, 3*(174–180), 207–217.

Joyce, C. (2009). New challenges in biogeography. *Area, 41*, 354–357.

Kinch, M. P. (1980). Geographical distribution and the origins of life: The development of early nineteenth century British explanations. *Journal of the History of Biology, 13*, 91–119.

Larson, J. (1986). Not without a plan: Geography and natural history in the late eighteenth-century. *Journal of the History of Biology, 19*, 447–488.

Larson, J. L. (1994). *Interpreting nature: The science of living form from Linnaeus to Kant*. Baltimore: Johns Hopkins University Press.

Llorente-Bousquets, J., Papavero, N., & Hernández, A. B. (2000). Síntesis histórica de la biogeografía. *Revista de la Academia Colombiana de Ciencias Exactas, Físicas y Naturales, 24*, 255–278.

Lomolino, M. V., Riddle, B. R., & Brown, J. H. (2010). *Biogeography*. Sunderland: Sinauer Associates Inc.

Luna-Vega, I., & Luna-Vega, I. (2008). Aplicaciones de la biogeografía histórica a la distribución de las plantas mexicanas. *Revista Mexicana de Biodiversidad, 79*, 217–241.

Mayhew, R. (2001). The effacement of early modern geography (c. 1600–1850): A historiographical essay. *Progress in Human Geography, 25*, 383–401.

Mayr, E. (1944). Wallace's line in the light of recent zoogeographic studies. *Quarterly Review of Biology, 19*, 1–14.

McCarthy, D. (2009). *Here be dragons: How the study of animal and plant distributions revolutionized our views of life and earth*. New York: Oxford University Press.

Merriam, C. H. (1892). The geographical distribution of life in North America with special reference to the Mammalia. *Proceedings of the Biological Society of Washington, 7*, 1–64.

Millington, A., Blumler, M., & Udo, S. (2011). *The SAGE handbook of biogeography*. Los Angeles: Sage Publications.

Morrone, J. J. (2009). *Evolutionary biogeography: An integrative approach with case studies*. New York: Columbia University Press.

Müller, G. H. (1992). Ratzel et la biogéographie en Allemagne dans la deuxième moitié du XIXe siècle. *Revue d'histoire des sciences, 45*, 435–452.

Müller, G. H. (1996). *Friedrich Ratzel (1844–1904): Naturwissenschaftler, Geograph, Gelehrter*. Stuttgart: GNT Verlag.

Nelson, G. (1978). From Candolle to Croizat: Comments on the history of biogeography. *Journal of the History of Biology, 11*, 269–305.

Nelson, G., & Platnick, N. I. (1981). *Systematics and biogeography: Cladistics and vicariance*. New York: Colombia University Press.

Nicolson, M. (1987). Alexander von Humboldt, Humboldtian science and the origins of the study of vegetation. *History of Science, 25*, 167–194.

Nordenskiöld, E. (1936). *The history of biology: A survey. Translated from the Swedish by Leonard Bucknall Eyre*. New York: Tudor.

Nordstrom, L. O. (1990). "Bradfords Law" and the relationship between ecology and biogeography. *Scientometrics, 18*, 193–203.

Nyhart, L. K. (2009). *Modern nature: The rise of the biological perspective in Germany*. Chicago: University of Chicago Press.
Padian, K. (2008). Darwin's enduring legacy. *Nature, 451*, 632–634.
Papavero, N., Teixeira, D. M., Llorente-Bousquets, J., Papavero, N., Teixeira, D. M., & Llorente-Bousquets, J. (1997). *História da Biogeografia no período pré-evolutivo*. São Paulo: Editora Plêiade and Fundação de Amparo à Pesquisa do Estado de São Paulo.
Papavero, N., Teixeira, D. M., & Prado, L. R. (2013). *História da Biogeografia: o Gênesis à primeira metade do século XIX* (Vol. 1). Rio de Janeiro: Technical Books.
Parenti, L. R., & Ebach, M. C. (2009). *Comparative biogeography: Discovering and classifying biogeographical patterns of a dynamic Earth*. Berkeley: University of California Press.
Platnick, N. I., & Nelson, G. J. (1978). A method of analysis for historical biogeography. *Systematic Zoology, 27*, 1–16.
Quammen, D. (1997). *The song of the dodo: Island biogeography in an age of extinctions*. New York: Scribner.
Ratzel, F. (1888). Die Anwendung des Begriffs "Oekumene" auf geographische Probleme der Gegenwart. *Königlich Sächsische Gesellschaft der Wissenschaften zu Leipzig Philologisch – Historische Klasse, 40*, 137–180.
Ratzel, F. (1891). *Anthropogeographie* (Vol. 2). Stuttgart: Engelhorn.
Raup, H. (1942). Trends in the development of geographic botany. *Annals of the Association of American Geographers, 32*, 319–354.
Reed, H. S. (1942). *A short history of the plant sciences*. Waltham: Chronica Botanica Company.
Rehbock, P. F. (1983). *The philosophical naturalists: Themes in early nineteenth-century British biology*. Madison: University of Wisconsin Press.
Rickleffs, R. E. (2008). *The economy of nature*. New York: W.H. Freeman and Company.
Riddle, B. R. (2005). Is biogeography emerging from its identity crisis? *Journal of Biogeography, 32*, 185–186.
Rübel, E. (1927). Ecology, plant geography, and geobotany; Their history and aim. *Botanical Gazette, 84*, 428–439.
Schmithüsen, J. (1970). *Geschichte der geographischen Wissenschaft von den ersten Anfängen bis zum Ende des 18. Jahrhunderts*. Zürich: Bibliographisches Institut.
Schmithüsen, J. (1985). Vor- un Frühgeschichte der Biogeographie. *Biogeographica, 20*, 1–166.
Spellerberg, I. F., & Sawyer, J. W. D. (1999). *An introduction to applied biogeography*. Cambridge: Cambridge University Press.
Stromeyer, F. (1800). *Commentatio inauguralis sistens historiae vegetabilium geographicae specimen*. Göttingen: Heinrich Dieterich.
Tobey, R. C. (1981). *Saving the prairies: The life cycle of the founding school of American plant ecology, 1985–1955*. Berkeley: University of California Press.
van Oosterzee, P. (1997). *Where worlds collide: The Wallace line*. Kew: Reed.
Verran, H. (2009). On assemblage. *Journal of Cultural Economy, 2*, 169–182.
von Hofsten, N. G. E. (1916). Zur älteren Geschichte des Diskontinuitätsproblems in der Biogeographie. *Zoologische Annalen Zeitschrift für Geschichte der Zoologie, 7*, 197–353.
Wallaschek, M. (2009–2013). *Fragmente zur Geschichte und Theorie der Zoogeographie* (9 vols.). Halle (Saale): Published privately.
Williams, D., & Ebach, M. C. (2008). *Foundations of systematics and biogeography*. New York: Springer.
Withers, C. W. J. (2006). Eighteenth-century geography: Texts, practises, sites. *Progress in Human Geography, 30*, 711–729.
Worster, D. (1977). *Nature's economy: The roots of ecology*. San Francisco: Sierra Club.
Wulff, E. V. (1932). *Introduction to the historical geography of plants. Bulletin of Applied Botany, Genetics and Plant Breeding*. Leningrad: Institut prikladnoĭ botaniki i novykh kul'tur.
Wulff, E. V. (1943). *An introduction to historical plant geography*. Waltham: The Chronica Botanica Company.

# Chapter 2
# Origins, Race & Distribution

## Buffon's Law in Eighteenth Century Natural Classification

> *"[Buffon] the great French naturalist caught sight at once of a general law in the geographical distribution of organic beings, namely the limitation of groups of distinct species to regions separated from the rest of the globe by certain natural barriers. It was, therefore, in a truly philosophical spirit that, replying on the clearness of the evidence obtained respecting the larger quadrupeds, he ventured to call in question the identifications announced by some contemporary naturalists of species of animals said to be common to the southern extremities of America and Africa"* (Lyell 1842, p. 112).

In a melancholic introduction to the *Institut d'Égypte*, Étienne Geoffroy Saint-Hilaire laments the two lost battles in the newly acquired Egyptian colony by the French Army on the 30th Ventôse, year 9 (March 21st, 1801). "At the moment when we were informed of our disasters, and when the report circulated of them immediately excited against us the whole population of Egypt, a crocodile was brought to me which had been carried alive to Cairo, and, which had died three days before" (Geoffroy Saint-Hilaire 1803a, p. 136).

Geoffroy Saint-Hilaire dissected the Nile crocodile (*Crocodylus niloticus*), "celebrated by antient [sic] authors" restricting his examination of the organs. What struck Geoffroy Saint-Hilaire was how similar the Nile crocodile was to the recently discovered crocodile of Saint Domingo (part of present-day Haiti) (Fig. 2.1).

The first specimen brought to Geoffroy Saint-Hilaire was that of the "captain-general Leclerc, being informed by some officers of his staff who had served in Egypt that the crocodile of Saint Domingo had a great resemblance to that of the Nile, thought it a matter of importance to furnish naturalists with the means of confirming this circumstance: he therefore was desirous of making a sacrifice to us of two crocodiles which had been presented to him" (Geoffroy Saint-Hilaire 1803b, pp. 233–234). The discovery of the same kind of crocodile, in both the Nile and

*Crocodile of S.t Domingo.*

**Fig. 2.1** The Crocodile of Saint Domingo *Crocodylus acutus* Cuvier 1807 (Pictured in Geoffroy Saint-Hilaire 1803b, Plate V)

in Saint Domingo, had gone against what many naturalist to have believed, namely that the New World had the same, or at least incredibly similar, species of crocodile living in a similar environment.

> Our first suspicion on receiving this animal was that the identity of species was proved, and that thus the real crocodile existed in the warm countries of both hemispheres.
> 
> This, however, was a result so contrary to one of the finest laws established by Buffon, a law of the greatest importance in zoology as well as in the history of the revolutions of the globe, that I did not think proper to admit this first idea without a more accurate examination.
> 
> This law, founded on an observation Buffon had made, that no species of the torrid zone had been primitively placed in both continents, had either never been contradicted, or had been so only by objections the weak foundation of which had been soon discovered (Geoffroy Saint-Hilaire 1803b, p. 234).

At the same time Geoffroy Saint-Hilaire, Humboldt and Bonpland made a similar discovery during their travels in South America (1799–1803). They had observed several crocodiles in the Orinoco and Madalena that had "a resemblance so surprising to the crocodile of the Nile, that it required a minute examination of its parts to prove the law of Buffon relative to the distribution of species between the tropical regions of the two continents, was not erroneous" (Humboldt and Bonpland 1829, pp. 293–294).

The differences between both crocodiles was enough for Georges Cuvier to describe it as the American Crocodile (*Crocodylus acutus*) in 1807, based on characters from the jaw, neck plates and climate when compared to the vulgar or common form (Cuvier 1807, p. 55). Regardless, the discovery was not enough to convince Geoffroy Saint-Hilaire to "consider the law established by Buffon as invalidated" (Geoffroy Saint-Hilaire 1803b, p. 235).

Buffon's Law was used as a way to distinguish the differences between species from different areas for the purposes of identification and classification. The law Buffon proposed was stated that newer areas would have degenerate specimens or races originating from an older stock. The discovery of the Saint Domingo crocodile however was unusual as it occurred in the New World and differed very little from the African Nile crocodile. As an exception to Buffon's Law, Geoffroy Saint-Hilaire postulated, "it would be necessary to have a more accurate knowledge of the changes which crocodiles may undergo at the different ages; that is to say, whether they are not subject to local influences which produce accidental variations and to obtain some information respecting their habits". So why uphold a law that in practise may not actually work?

Much of eighteenth century zoological taxonomy was a shambles. There were many systems of classification starting with seventeenth century works like John Ray's *Historia Plantarum* (Ray 1686), and to Buffon's contemporaries, like Adanson's *Familles naturelles des plantes* (Adanson 1763) and Jussieu's *Genera Plantarum* (Jussieu 1789) and no nomenclature to speak of. Taxonomies were done according to the author's whim, some adhering to Linnaeus' system of Orders, Genera and Species (Brisson 1756; Pennant 1783), while others simply listed genera in alphabetical order (Martini 1774), added in extra divisions like families (Klein 1751) or, ignored Linnaean classification all together and listed species according to their uses to man (de Buffon 1749–1789).

Rather than describe a species based on a list of anatomical characteristics that was linked to a Latinised name, Buffon in his multi-volumed *L'Histoire Naturelle, générale et particulière, avec la description du Cabinet du Roi* published between 1849 and 1789, proposed a rule or law that accounts for all characteristics of organisms, including the climate they inhabit (Fig. 2.2). A lion, for instance, lives in the hot savannah of the old world. The environment shapes the lion, both in form, size and temperament. If we were to move the lion to a colder mountainous environment, we would expect a change in temperament. The lion becomes less aggressive, it changes its diet, becomes smaller and changes colour. The same stock can produce different species based on the environments in which they live; lions live in hot savannahs, and mountain lions in the cooler mountainous regions. Buffon had seemingly solved the problem of identifying taxonomic diversity by using climate and environment.

> ... the general rule [law] which I intend to establish, and which seems to me to be our only certain guide to the knowledge of animals. This rule [law], which leads us to judge of them as much by climate and disposition as from figure and conformation [morphology], will seldom be found wrong, and it will enable us to avoid and discover a multitude of errors. If, for example, we mean to describe the hyæna of Arabia, we may safely affirm that it does not exist in Lapland; but we will not say with Brisson,[1] and some others, that the hyæna and the glutton [wolverine] are the same animal [...] But it is not my object at present to point out all the errors of nomenclators [Linnaean taxonomists]; my intention is solely to prove that their blunders would have been less had they paid some attention to the differences of

---

[1] Buffon is referring to *Voyez le Règne animal* (Brisson 1756) by Mathurin Jacques Brisson (1723–1806) the French naturalist and taxonomist.

# HISTOIRE
## NATURELLE,
### GÉNÉRALE ET PARTICULIÈRE,
*AVEC LA DESCRIPTION*
## DU CABINET DU ROI.

*Tome Sixième.*

## A PARIS,
### DE L'IMPRIMERIE ROYALE.
M. DCCLVI.

**Fig. 2.2** Frontispiece of the sixth edition of Buffon's *Histoire naturelle, générale et particulière, avec la description du Cabinet du Roi. Tome Sixième* (1756)

climates[2]; if the history of animals had been so far studied as to discover, which I have done, that those of the southern parts of each continent are never found in both; and lastly, if they had abstained from generic names, which have confounded together a number of species, not only different, but even remote from each other (Buffon 1807, pp. 50–51).[3]

Buffon's law was essentially no different from the writings of Aristotle, who proposed a similar rule.[4] Regardless, Buffon's law was an essential part of his *L'Histoire Naturelle,* or *Natural History,* which aimed to describe organisms by "climate and disposition as from figure and conformation" without the need for a hierarchical classification (i.e., "generic names"). In contrast, Linnaean classification relied solely on unique diagnostic features that were linked to a Latinised binomial name. Buffon, quite rightly, saw Linnaeus' classification as artificial (as did Linnaeus), seeing no need for Latin names when a single common word would do. After all, Buffon writes, "why introduce an unintelligible jargon, when we may be understood by pronouncing a simple name?" (Buffon 1761, p. 52). Also, Linnaeus' system offered nothing in the way of a knowledge of the organisms' temperament and their influencing factors such as diet and climate.[5] What bothered Buffon most however, was the proliferation of Linnaean names.

Buffon believed that the "true business of a nomenclator is not to enlarge his list, but to form retinal comparisons in order to contract it. Nothing can be more easy then, by pursuing all the authors on animals, and by selecting their names and phrases, to form a table which however will always be long, in proportion as the enquiry is superficial" (Buffon 1807, p. 51). Buffon was practical and sought to

---

[2] William Sharp Macleay also points this out: "The very naturalists – such as Buffon, Reaumur, and Bonnet – who despised scientific nomenclature, were obliged to attend to classification" (Macleay 1819, p. 10).

[3] The same quote translated above by Nelson (1978, pp. 275–276) from the original French (see Buffon 1761, Vol. 9, pp. 118–120) is used throughout the history of science literature (e.g., Browne 1983; Miracle 2008). Nelson's translation, however, only includes the first sentence of the above quote (translated in Buffon 1807), thereby leading many to misread Buffon's Law as a law of distribution, rather than a law of classification based on climate and disposition. Baker (2007) also makes this same mistake by summarising Buffon's Law as "Areas separated by natural barriers have distinct species" (Baker 2007, p. 206). However Baker (2007) does state in a footnote that his "... principal sources for the historical details of the following discussion are Nelson (1978) and Fichman (1977)" (Baker 2007, p. 206, footnote 31).

[4] Buffon's Law is not only similar but most likely derived from Aristotle: "In many places the climate will account for peculiarities; thus in Illyria, Thrace, and Epirus the ass is small, and in Gaul and in Scythia the ass is not found at all owing to the coldness of the climate of these countries. In Arabia the lizard is more than a cubit in length, and the mouse is much larger than our field-mouse, with its hind-legs a span long and its front legs the length of the first finger-joint [...] Locality will differentiate habits also; rugged highlands will not produce the same results as the soft lowlands. The animals of the highlands look fiercer and bolder, as is seen in the swine of Mount Athos; for a lowland boar is no match even for a mountain sow" (Aristotle, *Historia Animalium*, VIII, 28, 29).

[5] As Jacques Roger elegantly states "It was not classing morphologies but in systematizing our knowledge of living beings as they live, through comparing their physiologies, their 'habits', according to the climates in which they live [...] That was the true goal of the *Natural History*" (Roger 1997, p. 90 original emphasis).

simplify his *Natural History* to suit a naturalist as much as a layperson. By applying Buffon's Law there were fewer names to remember, compared to the order, class, genus and species names proposed by Linnaeus. Given "that in the whole known part of globe there are not above 200 species of quadrupeds, including among them 40 species of apes. To each of these, therefore, we had only to appropriate a name; and to retain 200 names. Why change terms merely to form classes? When a dozen animals are included under the name, for example, of *the Rabbit*, why is the Rabbit itself omitted, and must be sought for under the genus of *the Hare*?" (Buffon 1807, p. 52 original emphasis). Why indeed? Buffon's system served to know the whole organism, while Linnaeus' system offered nothing more than a simple identification tool, "a *general index* to natural history".

Buffon's *Natural History* was a synthesis that allowed naturalists to create non-hierarchical classifications based on a deep knowledge of the organisms under study and his subsequent law merely a basic rule of identification. If so, why then, did nineteenth century naturalists like Charles Lyell and Theodore Gill,[6] mistakenly believe Buffon's Law to be "a general law in the geographical distribution of organic beings, namely the limitation of groups of distinct species to regions separated from the rest of the globe by certain natural barriers"?[7]

## A Note on the Differing Themes of Distribution and the Roles of Naturalists

I would like to divert the attention of the reader to discuss the role of scientific themes. Distribution, it seems, means different things to different people during the late eighteenth and early nineteenth centuries. For Buffon it was an additional characteristic with which to identify and classify. For others, like de Candolle, however, distributions formed distinct units that themselves could be classified into regions. Similar distributional classifications for example, were also by used Prichard to distinguish between human migrations. Each version of distribution, as

---

[6]American vertebrate taxonomist, Theodore Gill (1837–1914) stated in his annual presidential address, delivered to the Third Anniversary Meeting of the Biological Society of Washington on January 19, 1883 that "It is Buffon who is to be credited with having first promulgated precise generalizations respecting the geographical distribution of animals. Buffon, in this respect, not only advanced much beyond his predecessors, but leaped at once to a position which some of the more pretentious naturalists of our own times have failed to attain" (Gill 1885, p. 1). Gill does not elaborate on the identity of these "pretentious naturalists".

[7]Even earlier naturalists like William Sharp Macleay didn't believe that Buffon's Law was relevant to explaining distributions: "It had, it is true, been already observed by Buffon, that the animals of the new world are different from those of the old; and various travellers had shown that the productions of different countries bear a character peculiar to each. But these were all rude and fortuitous observations, which had no view whatever to general consequences, or to the development of those laws by which it is now certain that the geographical distribution of organized matter was regulated at the creation" (Macleay 1819, p. 42).

it were, are attempts to conceptualise a scientific theme, that is a set of problems, questions or tasks that attempt to improve or revise a particular topic, such as natural classification, physiology, ontology for example. By scientific theme, however, I do not mean a transformation from a topographical to historical process as Browne suggests,[8] which harks back to the German historical tradition when there was a shift from "mechanist to organic patterns of thought" (Larson 1994, p. 2). Rather these scientific themes are based on the aims and goals of the naturalist as well as their skills, types of data (e.g., distributions of plant species) and materials (e.g., museum collections, observations made in the field) to which they have access. For example, early nineteenth century naturalists like Zimmermann or Willdenow, who were confined to a region within an academy, university or herbarium, may not have had the means or capital to travel round the world, as did the Prussian aristocrat Alexander von Humboldt. Therefore Humboldt was able to make many observations and measurements (as well as use the collections). However, unlike Zimmermann or Willdenow, Humboldt may not have had the necessary scientific training to plan and execute detailed taxonomic surveys, keys and monographs. Instead Humboldt produced a unique classification of plant forms, observed and recorded phenomena and produced a very different work from that of a conventional nineteenth century naturalist who specialised in plant taxonomy.

The aims and goals of the naturalist are also important when considering distribution. Are they asking questions about classification? If so, then explaining or identifying centres of distribution are not vital to a stable and accurate classification (as was the case with Zimmermann, Willdenow and Linnaeus). If, for instance, the naturalist is asking questions about the history of a species, as did Link and Prichard, then centres of distribution are of considerable interest, so are migration patterns and explaining the forms of migration. Considering that Zimmermann, Willdenow, Link and Prichard are often described as being taxonomists, plant and animal geographers, they did very different things due to their different themes.

Distribution, like organisms, can be studied under various themes. One such theme is origins. By the late seventeenth century, and the discoveries of the New World, and the Great Southern lands, theologians mostly studied centres of creation. After all, creation was a theological question. The Dutch theologian, and minister dismissed from the Reformed Church, Abraham van der Mijle (1563–1637, a.k.a. Milius or Mylius see Hooykaas 1956) sought to ask "[H]ow, and for what reason did men and animals arrive, especially by land, to the uninhabited parts of the world" (Mylius 1667, p. 4).[9] Similar questions were posited by other theologians like Jesuit priest Athanasius Kircher (1601–1680), "Where did the animals of the regions

---

[8] As well as Joel B. Hagen "Prior to about 1900 biogeography was primarily a descriptive activity closely related to taxonomy" (Hagen 1986, p. 197).

[9] Translated from the original Latin: "Quomodo, quaque ratione tam homines, quam animalia catera, prasertim terrestria, Orbis terrarum singulas partes inhabitatum pervenerint". Comments such as these had led Hofsten to hail Mylius as the "forerunner of centre of origin studies" (Hofsten 1916, p. 32).

and islands of the globe come from?" (Kircher 1675, p. 111).[10] Both Mylius and Kirchner made use of the distributional knowledge of the time to make assumptions about the migrations of organisms.

Determining origins were part of theological discussions, that were briefly entertained by Link, Linnaeus and Zimmermann, but did not form part of their work or over-riding elements in their themes. Rather it was *distribution*, what was known and how it helped shaped their understanding of organisms and their classifications, which drove eighteenth century plant and animal zoogeography. The main question was not where organisms come from, but more practically, what are organismal distributions and how do they help us classify life?

Rather than a move from topographical to historical processes or from mechanist to organic thought, nineteenth century plant and animal geography was about moving from one theme to the next based on the limitations of the naturalist. This can be best seen in the work of German zoologist and geographer[11] Eberhard August Wilhelm von Zimmermann, his influences and his legacy.

## Zimmermann's Legacy: From the *Specimen Zoologicae Geographicae* to Nineteenth Century Animal Geography

> *"Until then, Zimmermann was one of the first naturalists who discussed the distribution of animals on earth in the rank of species-specific, and therefore, taxonomic characteristics" (Feuerstein-Herz 2004, p. 261).*[12]

> *"In the 19th century travel came to be considered as a tool of geography" (Beck 1957, p. 1; translated in Beck 1983, p. 73).*[13]

The middle to late eighteenth century saw two distinct types of accounts about the living world, namely natural histories and travelogues. Natural histories hadn't changed much since the *Historia plantarum* of English naturalist John Ray (1686) and were a descriptive inventory of natural objects, such as rocks, acids, plants and animals, their physiological, chemical and behavioural attributes, and their uses. Ray also contributed the same histories to zoology, namely *Historia*

---

[10]Translated from the original "Quomodo Animalia in Universas Globi Terreni Regiones et Insulas devenerint".

[11]Zimmermann is often referred to as a "Professor of Mathematics" (see Pennant 1783, p. ix, Bodenheimer 1955, p. 351, Schmithüsen 1985, p. 65, Bradley 2006, p. 17).

[12]The original reads "Bis dahin war Zimmermann einer der ersten Naturforscher, die die Verbreitung der Tiere auf der Erde im Rang eines arteigenen und damit taxonomischen Merkmals diskutierten" (Feuerstein-Herz 2004, p. 261).

[13]Beck's footnote reads: "By travel the author mean all enterprises that have enlarged our knowledge of the earth's surface or are still enlarging it at the present time and not just discovery or research expedition. Tourists, travel writers, and poets have also conducted journeys with geographical effects and should thus be considered in the history of travel" (Beck 1983, p. 99).

*insectorum* (Ray 1710) and *De Historia Piscium* (Willughby 1686). These natural histories simply grouped organisms based on their characteristics, a classification that Linnaeus systematised in his *Systema Naturae* (Linnaeus 1753). Linnaeus also included humans in his *Systema Naturae*, which naturalist left out of their natural histories, such as *Die Naturgeschichte der Thiere in sistematischer Ordnung* by Johann Samuel Halle (1757), *Regnum Animale in Classes IX Distributum* by Mathurin-Jacques Brisson (1756). Others, however, ignored humans altogether, namely in the *Quadrupedum dispositio brevisque historia naturalis* by Jacob Theodor Klein (1751), *Abbildungen nach der Natur mit Beschreibungen* by Johann Christian Daniel von Schreber (1774–1804) and *History of Quadrupeds* by Welsh naturalist Thomas Pennant (1783).[14]

Natural histories that adopted Linnaeus' approach included a name, a short diagnosis (in Latin), and a synonymy (when needed), either as a footnote or as a short introduction. Most of the text was a description of what was known about the animal, with a discussion about the taxonomic placements of previous authors and a comparison between known species. The geographical content however varied. Histories were limited to where species have been known to exist and collections sites. Other natural histories concentrated on the plants and/or animals of a particular area such as the *Flora Lapponica* of Linnaeus (1737), the *Flora Rossica* of Peter Simon Pallas (1784–1788), *Flora sibirica* of Johann Georg Gmelin (1747), and the *Histoire naturelle et civile du royaume de Siam* of François Henri Turpin (1771).

Natural histories are the forerunners to nineteenth century taxonomies, the names of which are still valid today. Travelogues, however, are mainly about organismal distributions within a given area, providing accounts of the characteristics of a geographical area, such as habitat, geography and climate. When compared to the natural histories above, the travelogues of the eighteenth century are far broader in scope; some including anthropological accounts, but generally lacks the systematic nature of dedicated natural histories. Travelogues were, after all, accounts of someone's travels. Some travelogues did incorporate natural histories, as in the *Histoire Naturelle du Sénégal* by Michel Adanson (1757). However, most were accounts of a voyage and places, like *Caput Bonae Spei Hodiernum* by Peter Kolbe (1719), the *Voyage de Bougainville, capitaine de vaisseau, autour du monde* of Louis Antoine de Bougainville (1771), or *Observations Made during a Voyage round the World* of Johann Reinhold Forster (1778). While there are differences between travelogues and natural histories, it is the observations made in travelogues that provide further knowledge about the distribution of known organisms. The true genius in Zimmermann's work is how he managed to meld natural histories and

---

[14]Pennant rejected Linnaeus' arrangement of the mammals to the extent that he was obliged "to separate myself, in this instance, from this crowd of votaries; but that my reflection may not appear effect of whim or envy, it is to be hoped that the following objections will have their weight. I reject his first division, which calls *Primates*, or Chiefs of the Creation; because my vanity will not suffer me to rank mankind with *Apes, Monkeys, Maucaucos*, and *Bats*, the companions Linnaeus has allotted us even in his last system" (Pennant 1783, pp. iii–iv, original italics).

travelogues. Rather than producing a detailed systematic catalogue of the different types of quadrupeds, Zimmermann produced a classification of where they are found.

When it was published, *Specimen Zoologicae Geographicae* was the only work that categorised the planet into known quadruped distributions, namely:

Chapter I: Animals dispersed throughout the world and their degeneration.
Chapter II: Introduction
Part One. Quadrupeds of both the Old and New World
Part the Latter: Quadrupeds of the Old World
Chapter III. Quadrupeds of the New World
Chapter IV. In which the animals are generally treated by the dispersion across the surface, whose consequences are added in the history of the planet (Zimmermann 1777, p. xxiv).[15]

Zimmermann's *Specimen* was not a strict natural history, with a hierarchy of orders and families, but a distribution of genera and species, just like one would except to see on a late nineteenth century map. In fact, Zimmermann published *Tabula mundi geographico zoologica sistens quadrupedes hucusque notos sedibus suis adscriptos* (Zimmermann 1777), possibly the first distributional map that gave certain Linnaean names and their distributions (Fig. 2.3).

Zimmermann's *Specimen* was built on both the travelogues and natural histories of seventeenth and eighteenth century naturalists. No single individual could travel the world in order to gather the distributional data for all known taxa. Naturalists were limited by the amount of funds, time and commitments and many couldn't afford the expeditions of Pallas, Adanson and Gmelin. So too was Zimmermann who relied on funds raised from writing prefaces to books and translations to undertake even the most minor travels abroad (see Feuerstein-Herz 2004 for a detailed account). Regardless, his *Specimen Zoologicae Geographicae* was solely reliant on the observations of others. But not everything went to plan.

A year after his self-described "Latin zoology" was published, an updated German version, titled *Geographische Geschichte,* appeared. The revised version was due to "recently published texts of Lord Kaimes [sic] [1774],[16] Professors Blumenbach [1775],[17] Schreber [1774],[18] Erxleben [1777], Kant [1775] and others"

---

[15]The original reads: "Caput I. De animalibus per totum pene terrarum orbem dispersis, eorumque degenerationibus; Caput II. Introductio; Pars Prior. De quadrupedibus magnos, tam antiqui quam novi orbis, tractus tenentibus; Pars posterior. De animantibus magnos antiqui tantum orbis tractus tenentibus; Caput III. De quadrupedibus arctioribus plagis novi antiquive mundi alligatis; Caput IV. In quo generatim de quadrupedum per teluris superficiem dispersione agitur, cui consectaria nonnulla telluris bistoriam illustrantia adduntur" (Zimmermann 1777, p. xxiv).

[16]Henry Holme, Lord Kames (1696–1782), Scottish judge and philosopher, published his *Sketches of the History of Man* in 1774 (Kames 1774).

[17]Blumenbach's doctoral thesis *De generis humani varietate nativa* (Blumenbach 1775, University of Göttingen).

[18]Zimmermann (1777, p. 450 footnote z) actually did see the first volume of Schreber's *Säugthiere* (1774).

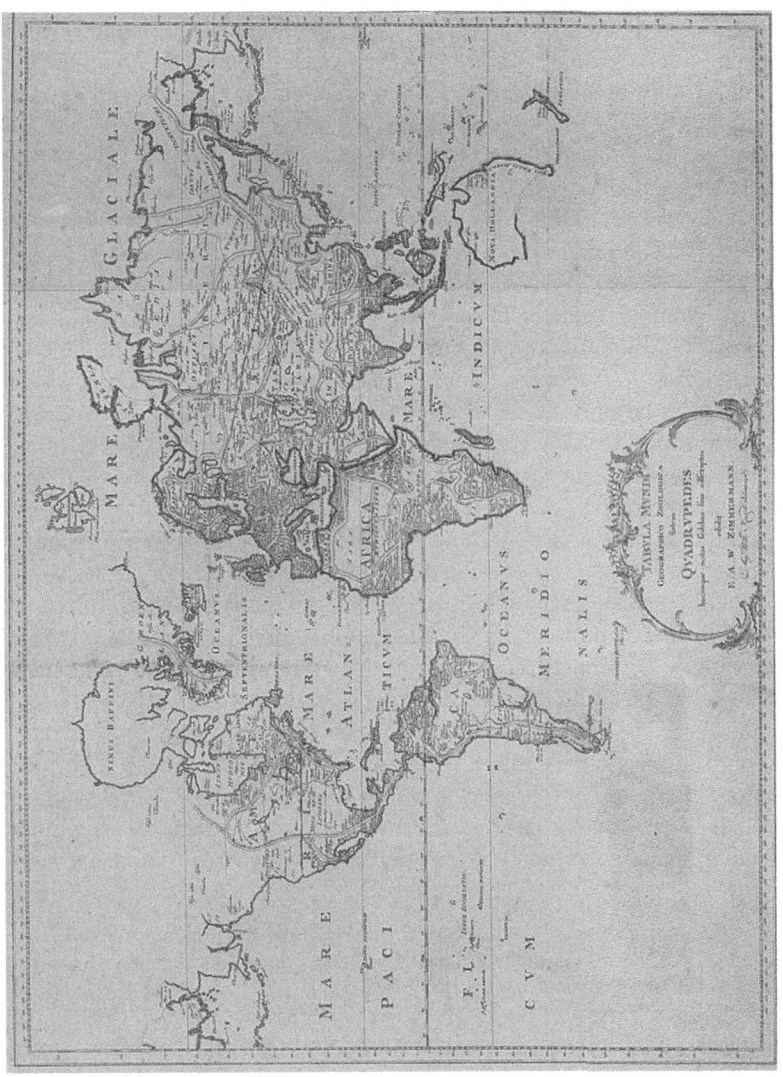

**Fig. 2.3** Zimmermann's *Tabula mundi geographico zoologica sistens quadrupedes hucusque notos sedibus suis adscriptos*, second edition of 1783. The revised map in the 1783 German edition contains the newly discovered Sandwich Islands (Hawai'i) and the Seychelles, which were absent in the original 1777 edition. However, while the German 1783 edition referred to the "Zoologische Weltcharte", the map remained in Latin in both editions (see Feuerstein-Herz 2004, p. 232, footnote 840) (Source: National Library of Australia)

that appeared during the 4 year delay of the printing of his text (ca. 1773). Zimmermann wasn't exactly pipped to the post by new data. Rather, a reliance on new editions meant more revisions, which on reflection, is far cheaper alternative than organising a rival expedition.

The *Specimen Zoologicae Geographicae* is a reflection of an eighteenth century naturalist's limitations. The text is proof that large expeditions to foreign lands are not necessary to pursue a scientific theme. Rather it makes animal geography available to everyone with access to natural histories and travelogues. The eighteenth century naturalist could therefore pursue a scientific theme (i.e., plant and animal geography) without the need for building collections or making observations.

Zimmermann's three volume *Geographische Geschichte* appeared in 1778, 1780 and 1783 respectively.[19] The work, similar to that of the *Specimen Zoologicae Geographicae* had a whole volume dedicated to human natural history, possibly due to the appearance of Kant's *Über die verschiedenen Rassen der Menschen* (Kant 1775), which Zimmermann mentions in the introduction to his first volume. The second and third volumes roughly follow that of the 1777 "Latin zoology", but with species ordered differently. In Chap. 2 (Part 1), "Quadrupeds of both the Old and New World" of the 1777 volume, Zimmermann lists the first 10 sections as 2. Reindeer and Elk, 3. Martens, 4. Beavers and otters, 5. The Lynx, 6. Wolverines, 7. Squirrels, 8. Marmots, 9. Badgers, and 10. The Water vole. In the German 1780 second edition, the same chapter title list the first ten sections as 1. the Lemming, 2. the Red-backed Vole, 3. European Ground Squirrel, 4. the Hamster, 5. the Brown Rat, 6. Common Shrew, 7. Water Shrew, 8. Wood Mouse, 9. Tundra Vole and 10. Siberian Jerboa. Many of the taxa of the German edition are missing in the Latin zoology once again emphasising the reliance on new natural histories and updated taxonomies. Zimmermann seemingly revised his *Specimen Zoologicae Geographicae* with the addition of a new volume of on human geography. Why then combine human geography with that of animal geography?

Zimmermann (1777) differed from Buffon in his seemingly polygenetic approach.[20] For Buffon, animals suited to warmer climes moved from a former warmer climate at the poles to the areas that they live in now. For Zimmermann, however, animals were created where they were found – their distributions fluctuating with the changes in climate, some species becoming extinct with extreme changes:

> When Zimmermann [...] discusses migrations, it must be clear that this term includes expansion of area [...] Natural migration is then the spread of animals because of a natural increase in their former area (Bodenheimer 1955. p. 355).

---

[19]For latter twentieth century American zoologist Joel Asaph Allen, *Zoologiae geographiae* was "constructed on nearly the same general plan [*Geographiche Geschischte*], however, is not merely a German translation [...] but an essentially different work" (Allen 1902, p. 13). There are few reviews of Zimmermann's work.

[20]For Zimmermann, "it was more rational to believe every animal was created in the area where it now lives ..." (Browne 1983, p. 26, also see Larson 1994, p. 83).

The only exception is humans, which appeared to live everywhere. Given the global distribution of humans, Zimmermann placed them into Chap. 1 (see above). Moreover, humans are treated like any other quadruped within the *Geographische Geschichte*, with detailed descriptions and comparisons between the different types of humans, where they occur, the climates they live in, their behaviour, sexual reproduction and nutrition. What sets his work apart from that of his Linnaean contemporaries is his classification of inter-breeding human races, rather than species, possibly explaining why he refers to the "recently published texts of Lord Kaimes [sic], Professors Blumenbach [1775], Schreber [1774], Erxleben [1777], Kant [1775]", all of which refer to the races of humans. A geography that describes inter-breeding races of a widespread species, which had adapted to all climates, is a far cry from a polygenetist approach.

> An enquiry here presents itself to no small moment: are the human beings which possess these various and opposite powers, derived from one common origin: or were different races of men formed and adapted by their original make to their specific climates? M. Zimmermann enters fully into this question, which has so frequently agitated (Anon. 1789, p. 683).

Since Zimmermann clearly entertained the idea of a single origin for humans, then where does the notion of polygenesis derive from? Perhaps it derives from Zimmermann's reading of Lord Kames or Kant, or perhaps even in the rejection of a Biblical centre of creation,

> Apart from the distribution of animals and plants, Linnaeus appears to have lifted these great problems [of a Biblical centre of creation]. It is not clear how such a man could not notice the impossible descent of animals from these mountains [Ararat] (Zimmermann 1783, p. 194, my translation).[21]

The notion that Zimmermann had a reason for variation, beyond that of inter-breeding races, is one that comes from a purely modern evolutionary view-point. Zimmermann was not interested in explaining the distribution of animals. In Hofsten's history of disjunction, he bemoans the fact that Zimmermann's interest lay elsewhere; rather he "didn't seek to explain the distribution of vicariant species. This might be a weakness, but we can not forget, that the time was not ripe for solving these questions, and that these limitations allowed him to gain the valuable results that he did" (Hofsten 1916, p. 59, see also Larson 1994, p. 103).

Our present day interests may not match those of eighteenth century naturalists. While they may have been on the verge of "discovering" some form of evolutionary theory, they were not "evolutionists". Their interests were mainly in understanding what these species and races were and where they were distributed.

Eighteenth century scientists quite rightly saw any explanation of variation beyond that of inter-breeding, as it is purely hypothetical, once the bread and butter

---

[21] The original reads: "Glücklich scheint Linné die größten Schwierigkeiten bis auf das weitere Verbreiten der Thiere und Pflanzen geben zu haben; allein es ist kaum begreiflich, wie einem solchen Manne die Unmöglichkeit des herabsteigend der Thiere von diesem Gebirge nicht auffallen mußte" (Zimmermann 1783, p. 194).

of theologians and not scientists. While such musings may dominate the prefaces of natural histories, they are themselves not the main feature of the work. Historians of science have occupied themselves with explaining these hypothetical ideas (e.g., Bodenheimer 1955; Larson 1994; Browne 1983), but these were not the main pre-occupations of the naturalist. Sorting through classifications and understanding what occurs where, are far more constructive pursuits as they make sense of the data available at the time. The task Zimmermann and his contemporaries had set themselves was to understand these classifications and determine their stability. Merely hypothesising the origins of a species lacked evidence and did not itself make a scientific theme. However, asking whether a species was a natural entity was a far more intellectual pursuit. But what of Zimmermann's legacy?

Larson suggested that "*Geographische Geschichte* was an influential model for this more modest science.[22] Many naturalists accepted Zimmermann's strategy; they avoided fruitless speculation and contented themselves with attainable goals" (Larson 1994, p. 103).[23] These "attainable goals" of a "modest science" were a result of reading literature rather than of travelling, observing and collecting. Anyone with access to the right literature could, as Larson claims, use this "influential model". The first to do so was, unexpectedly, a German botanist and chemist, Friedrich Stromeyer.

A student of Gmelin at the University of Göttingen, Stromeyer completed his doctoral thesis on plant geography in 1800, which was published as *Historiae Vegetablium Geographiae Specimen* (see Chap. 3 for a detailed account). Stromeyer was unusual in his adoption of Zimmermann's system, categorising plant geography into: Phytogeography, the known distribution of plants and the laws of distribution, which he termed; Historical plant geography, namely the history of present day plant distributions, which he termed and; Applied geographical historical botany, that is, the relation between human and plant distributions. Like Zimmermann, Stromeyer

---

[22]From the perspective of a twenty-first century scientist Larson's reference to a "more modest science" is puzzling. The "modest science" Larson is referring to was in need of "more accurate, better-coordinated information on physical conditions and actual distribution on a worldwide scale. With the inchoate information naturalists possessed, they could only speculate aimlessly about possible migrations, adaption to new habitations, and earlier geographical conditions" (Larson 1994, p. 103). The actual distributions for most species are still relatively unknown; most information we have on earlier geographical conditions are still poor and possible migrations and adaptions to new habitations are still hotly debated. The geographical studies of the twenty-first century science, like that of the eighteenth, are modest.

[23]Compare this to the history of Gunnar Broberg, in which Zimmermann's work is summed up by his law that states the sum of "types of organised bodies increases with the degree of sensation of life (Broberg 1990, p. 60)". For Broberg, Zimmerman's "account did not visibly change the direction of natural history, but it forced two themes to surface: the dramatic growth of numbers and the pessimistic prospects for complete knowledge" (Broberg 1990, p. 60). Why Broberg chose to look at a small part of Zimmermann's work in order to assess its impact on eighteenth century geography could possibly be related to the topic of the book in which Broberg's chapter is published: *The Quantifying Spirit in the Eighteenth Century.*

divided up the plant world geographically, not by their classes or orders, but very much in the style of Zimmermann,

> After the example of the celebrated Zimmermann [1777; 1778–1783], who so excellently published on a similar subject in the matter of animals, I believe that I can gather the whole outline of this subject, so full and abundant, scarcely inconveniently under this common notion, and encompass it by the denomination of the Geographic History of Vegetables (Stromeyer 1800, pp. 14–15)

Stromeyer set out a system for plant geography, similar to how Zimmermann practised his geography, by examining the published literature, namely existing histories of plant geography already published, commentaries, and floras, travelogues and topographies, as well as the geographies of single taxa and general geographies of humans. Stromeyer's system borrows from Zimmermann by cataloguing known plant species into geographical categories.[24]

The system proposed by Stromeyer really set the scene for plant geography. For instance, prior to his Stromeyer's *Specimen*, plant geography was presented as a history within large botanical syntheses, usually at the end as in the case of Linnaeus or Willdenow. In other works, such as Humboldt's *Flora Fribergensis* (Humboldt 1793), plant geography was bemoaned as being confused with plant history, certainly true in the case of Carl Ludwig Willdenow's *Grundriss der Kräuterkunde* (Willdenow 1792), the bible for eighteenth century botany. Apart from the sections on explanations for distribution in Linnaeus, Link, Giraud Soulavie and Willdenow (as well as the minor commentary by Humboldt), there was little written on plant geography, a field that began to flourish in the early nineteenth century.[25]

Stromeyer's proposed system of cataloguing species based on geography is modern, resembling a nineteenth century plant geography, which relies heavily on the published literature and focuses away from check lists or strict taxonomies, like that of his contemporaries. Unfortunately, Stromeyer was forgotten in nineteenth and twentieth century botany, with only a handful of practitioners referring to his work (see Chap. 3), possibly because he did not propose any new syntheses or ideas nor did he follow up and publish the rest of his thesis.[26] With no real place in the history of ideas, Stromeyer, however, does have a place in the history of *practicing* plant geography, mostly as a revisional monographic framework.

Plant and animal geographers had found a system for cataloguing species, but it lacked consistency. For example, what sort of geography does one use to catalogue life? Zimmermann used the New and Old Worlds. Stromeyer chose to order plants according to varieties of different habitats, numbers and different sized ranges. However neither geographies were comparable. Given that many, if not all, plant

---

[24]These include namely the present distribution of land and aquatic plants, comparison between plant and animal distributions, past distribution of plants and fossil distributions.

[25]For a complete list of eighteenth century botanists writing on plant geography, see the translation of Stromeyer's introduction in the Appendix.

[26]Stromeyer only published the first chapter of the first book of his planned work. See Chap. 3 for a detailed account of Stromeyer's work.

and animal geographers were in fact comparative anatomists, the lack of a common geography would have been galling. It is no wonder then that the early nineteenth century was a period of regionalisation, beginning with Augustin de Candolle and James Cowles Prichard and ending with Alfred Russel Wallace and Clinton Hart Merriam (see Chap. 5). Plant and animal geographers of the nineteenth century also carried on the tradition of proposing distributional laws and measuring natural phenomena in order to uncover potential patterns.

## Distributional Laws

The relationship between distributional laws and the classification of geography is complex. The main challenge by Alexander von Humboldt was to keep history and geography as two separate entities. History, Humboldt believed, could not be measured as accurately as present day phenomena. "Earth history, more closely affiliated with geography than with nature study", Humboldt said, were not part of plant geography and "zoological history, the history of plants, and the history of rocks, which tell only the past state of the earth, are to be clearly distinguished from geography". Although the statement appears contradictory, Humboldt does have a point. Why involve hypotheses of what may have happened when we have a whole world in front of us. Surely present day organic and inorganic factors are driving modern day distributions?

Augustin Pyramus de Candolle had independently come across the same idea. Temperature, modes of watering and soil mobility were factors that he thought drove distribution. Like Humboldt, de Candolle sought quantifiable laws of distribution, laws that included inorganic factors that were equally comparable across different environments, like temperature, hydrology and soils. The main problem was that inorganic factors were not comparable. Temperature was not constant at the same latitude across the globe, nor was soils similar in similar environments. While this posed a huge problem for de Candolle, who was trained as comparative anatomist, it did not bother Humboldt, a generalist who was not confined to Linnaean taxonomy and comparative anatomy. While de Candolle finally gave up on soils and hydrology, he attempted to show that temperature is linked to latitude, a method that was not adopted, rather rejected as topographical (that is, descriptive) by his son Alphonse.[27] Humboldt however, managed to find a comparative link – vegetation types or plant

---

[27]Browne (1983, pp. 84–85) suggests that topographical, or statistical plant geography, was atheoretical and devoid of process. Browne's division between topographical (descriptive botanical arithmetic) and a historical (explanatory processes over time) phytogeography clearly has no bearing on what was practiced at the time. A.P. de Candolle, for instance, did use statistics in the reasoning behind his regionalisation of the world, but at the same time pondered which climatic or soil processes were responsible for these plant regions. The dichotomy between "science as a pattern" versus "science is a process" is artificial and says more about the thinking of the history and philosophy of scientists during the 1980s (*sensu* Hull 1988) than it does about the way science

forms – that were consistent across similar environments across the world. Ironic as it may seem, namely a non-comparative naturalist like Humboldt discovering the key for comparative plant geography, de Candolle started plant regionalisation, a first step toward a comparative plant geography. Botanists that used the same sets of areas were able to compare the different types of plants in each geographical region. De Candolle's method (botanical regionalisation) was different to that of Humboldt (classification of plant forms and vegetation), the former appealing more to taxonomists (see Chap. 5). Regionalisation in animal geography however has a different history.

The first attempt at a regionalisation was by Prichard (1826). Animal geographers were primarily interested in the migrations of animals. The zoogeographical map of Zimmermann (1777, 1783) for example, shows the present day location of quadrupeds, with lines denoting the northern and southern extents of certain species. While the map was to show the distribution of quadrupeds, it also attempted to incorporate a distributional law in it as well, namely the climatic isothermal lines that traced the northern and southern boundaries of animals. Zimmermann however used a modern geographical map[28] and did not propose zoogeographical regions.[29] In his second edition of *Researches into the Physical History of Mankind*, Prichard proposed several zoogeographical regions of quadrupeds:

> Hence by a reference to the geographical site of countries, we may divide the earth into a certain number of regions, fitted to become the abodes of particular groupes [sic] of animals; and we shall find on inquiry, that each of these provinces, thus conjecturally marked out, is actually inhabited by a distinct nation of quadrupeds, if we may use that term (Prichard 1826, p. 54).

Prichard defined zoogeographical regions by geographical barriers and the animals that lived in them. Moreover, these regions only applied to quadrupeds and not humans, insects, marine creatures and so on. How then were all animals and plants compared? Was there a common law of distribution?

---

was practiced in the mid nineteenth century. Browne's classification is dismissed herein (see also Maroske 2012).

[28]Zimmermann cites the sources of his 1783 zoological atlas [zoologichen Weltcharte]: "The mountain chains, some of which are found on the map, are not simply taken from [Phillippe] Buachen's Mountain Map [Buachen 1757], but are updated partly from the D'Anville maps [Jean Baptiste Bourguignon d'Anville 1697–1782] and, from the new observations of [Peter Simon] Pallas and [Ivan] Islenev [and Chudjakov, Efim Maksimovič 1777]" (Zimmermann 1783, p. 5). For a detailed history Buachen's maps see Debarbieux (2008).

[29]Rather, Zimmermann was interested in showing the living space [Wohnplatz]: "This attempt at a zoological atlas not only shows how many quadrupeds are current known, but also their current Wohnplatz" (Zimmermann 1783, p. 3, my translation) (see Chap. 5).

## Climate as a Law of Distribution

Possibly the first general law of distribution has been climate. Since Buffon proposed animals to be classified by their form and the environments that they live in, climate has been viewed as an obvious barrier for distribution. The first to quantify and map climatic factors, such as barometric pressure, temperature and altitude was Humboldt and Bonpland (1807, 2009) during their expedition to Latin America (1799–1804). The measurements made by Humboldt and Bonpland were incredibly detailed, resulting in the famous *Tableau physique*, a cross section of Chimborazo, indicating the distribution of plants against geographical phenomena, such as altitude, temperature, electrical phenomena and soil type[30] (Fig. 2.4). Moreover, Humboldt was aware that his method had limitations within a comparative framework. While you could compare the Andes with Alps based on geographical features such as altitude and temperature, there was no indication of how you would distinguish the features between "African plants and those of the New World? What are the analogies in shape that link the alpine plants of the Andes with those of the high peaks of the Pyrenees? These questions" Humboldt declares, "have hardly been debated till now and are without a doubt worthy of the physicist's attention" (Humboldt and Bonpland 2009, p. 73). In order to remedy this, Humboldt did something remarkable. Rather than list plants by the regions in which they occur (as was proposed by Stromeyer), Humboldt created new categories "which can be arranged into families or groups that are more or less analogous to each other". Humboldt names 15 groups, including "the palms, the tree ferns, the pines, heaths, the orchids", and in doing so effectively breaks away from conventional Linnaean taxonomy and creates a new classification of vegetation types (see Chap. 4 for a more detailed discussion). Dividing the world into such vegetation types or plant forms similar to Buffon's Law, but with a difference. The similarity lies in that the vegetation types or forms are governed by climate as well as morphology. However, unlike Buffon's Law these taxa have worldwide distributions, and do not require a common stem species or form. In other words, it was Humboldt and not Buffon, who had founded the first distributional law.

Zoologists did not convert Humboldt's law to a similar system in zoogeography. Rather zoologists like Julius Minding, Edward Forbes, James D. Dana, and Constantin W. Lambert Gloger continued to use isothermal lines, in the manner of Zimmermann, to denote the distribution of animals (Minding 1829; Gloger 1833; Forbes 1846; Dana 1853).[31]

---

[30] German polymath, Johann Wolfgang von Goethe was the first to use Humboldt's method to compare the Alps and Andes (Goethe 1813). Jackson (2009, p. 47) provides a summary of the work (see Chap. 5).

[31] So too did English botanist John Barton (1836–1908), who preferred to stick to Linnaean names and not employ the use of vegetation types or plant forms.

**Fig. 2.4** Humboldt's *Tableau physique* showing a cross section of Mount Chimborazo and Mount Cotopaxi in the Andes. The full title of the map reads *Geographie des plantes equinoxiales: tableau physique des Andes et Pays voisins. Dressé d'après des observations et des Mesures prises sur les lieux depuis le 10.ᵉ degré l'attitude australe en 1799, 1800, 1801, 1802 et 1803* (in Humboldt and Bonpland 1807) (Source: http://cybergeo.revues.org/docannexe/image/25478/img-7.jpg)

## *Vegetation and Faunas as a Law of Distribution*

Humboldt's vegetation types and plant forms were initially taken up by plant geographers, creating a duality within botany between the method of Humboldt and that of de Candolle. Humboldt's plant geography, however, was far more practical as it described natural forms of vegetation that are directly influenced by climate. For example, heaths may refer to a variety of small shrubby bushes that live on acidic soils, often sandy and well drained. As a heath defines a type of vegetation type and the types of species that are found within it, two heaths into different places may share no species in common. Compare this to a taxonomic region, which is defined by species distribution. In de Candolle's method, however, the distribution of a genus depends on a good understanding of the taxonomy. Change the taxonomy and you immediately alter the distribution, effectively making the distribution of poorly known groups highly unstable. The appeal of Humboldt's method is that accurate taxonomies will not affect the distribution of vegetation types. An added bonus is that the distributions of heaths, for instance, don't need to be explained via the dispersion of a stem species. Distribution is effectively a result of climate and surrounding geography and due to any centre of creation. Compare Franz Julius Ferdinand Meyen, an earlier adopter of Humboldt's method with the approach of zoologist William Swainson:

It is very easy to show that conditions of climate particularly heat and moisture, are the chief causes which determine the station and distribution of plants; and therefore it is of the greatest importance to the science of botanical geography, to know exactly the modes in which the influence of the often extremely complicated conditions of climate becomes apparent (Meyen 1846, p. 8).

That the primary causes which have led to different regions of the earth being peopled by different races of animals, and the laws by which their dispersion is regulated, must for ever hid from human research (Swainson 1835, p. 9).

Clearly Humboldt's method offered a far better law of distribution than that of the taxonomists and their natural regions. By the end of the nineteenth century, zoologists had an analogy to vegetation in the form of life zones:

The principles of geographic distribution of terrestrial animals and plants in the Northern hemisphere were clearly recognized in 1889; the correlation of the life zones was completed in 1892; the laws of temperature control were formulated in 1894 (Merriam 1894 [1895], p. 213).

In his 1826 *Essay on Botanical Geography,* Schouw attempts to define phyto-geographical regions "1. that at least half of the species should be peculiar; 2. That at least a quarter of the genera should be proper to the region [ ... ] 3. that individual families of plants be either peculiar to the region or else have their maxima" (Schouw 1823, p. 164 original italics). The regions and provinces in Schouw's view "are named after the vegetable forms that characterise them". This form of area classification marks the beginnings of a flora or vegetative regionalisation, particularly when "the phyto-geographic divisions of the globe, the boundaries and circuit of each, its climatic and other physical relations, should be described [ ... ] and lastly, a view should be given of the whole habit and character of the vegetation" (Schouw 1823, p. 164). Schouw's regionalisation is one in which a law has been proposed, namely vegetation and their corresponding taxonomies. Not surprising, Schouw's area classification are based on climatic zones (e.g., Flora Aplino-arctica) and on the dominant vegetation type (e.g., Provinca *Cichoriacearum*).

## *Regionalisation and the Law of Distributions*

English Zoologist William Swainson writing in his *Treatise on the geography and classification of animals* hails Danish zoologist Johann Christian Fabricius (1745–1808) as "the first naturalist who ventured on any actual definitions of he conceived to be natural climates or provinces, and his views are confined to the insect world" (Swainson 1835, p. 10). The concept of "climate particularly heat and moisture [to be] the chief causes which determine the station and distribution of plants" underpins the notion of natural regions or provinces. Animals do not form vegetation types like plants, which may be compared across different regions. Instead natural regions can be defined by naturally occurring species without having to create a new classification that departs dramatically from well-established Linnaean system.

Natural regions defined by taxa also appeal to taxonomists, plant and animal geographers who wish to catalogue their taxa by region. The disadvantage in using natural regions is that not all taxa fall into them neatly. By the 1850s, there were already many different competing regions for vertebrates alone. Taxonomists working at smaller scales had multiple distribution maps of animals that overlapped with different regions. The compartmentalisation of zoology (as opposed to botany) meant that different animal groups had different regionalisations. Forbes (1856) created regions of marine life based on "homoiozoic belts", effectively along isothermal lines. The proliferation of regionalisation in both plant and animal geography was an indication that natural regions differed significantly for different organisms. Debates as to the correct classification of the Nearctic and Palaearctic regions broke out in the pages of *Nature*, with Alfred Russel Wallace's vertebrate regions at loggerheads with those of Alfred Newton, Theodore Gill and Angelo Heilprin (see Chap. 5). Natural regions lacked a law of distribution that was common to all organisms. Regionalisation in botany had a similar fate. The first regions proposed by August de Candolle (1805, 1820) were dismissed by his son Alphonse de Candolle (1855) as "artificial systems" are a detriment to science "when they considered to be natural" (de Candolle 1855, pp. 1304–1305).

Swainson may have been correct in stating that laws of distribution were indeed hidden from the nineteenth century plant and animal geographer, however, the distributional laws governing individuals species may be obvious. Charles Lyell in his *Principles* points out the physiological attributes of plants and animals, such as winged seeds as potential mechanisms for dispersion for plants, and local geographical phenomenon like rivers and currents. While individual accounts of dispersion, whether by physiological means, by land bridges or through the aid of birds, *ad hoc* dispersion does not account for natural regions. In proposing possible mechanisms for distribution, Lyell like many others in the nineteenth and twentieth centuries, move further away from a common law and towards a multitude of causes rather than a set of laws that define natural regions, similar to Schouw's phytogeographical provinces. For Lyell, it is distribution of "original stocks" over "the whole surface of land and water, there would be never-the-less arise distinct botanical and zoological provinces, for there are a great many barriers which oppose common obstacles to the advance of a variety of species" (Lyell 1842, p. 167). If this is the case, why mention distributional mechanisms at all? The reason is that the distribution patterns of species, genera and vegetation, were small enough to be explained by individual laws, while larger regionalisation of vertebrates or land plants required larger mechanisms that were not viable or reasonable at the time. In any case, nineteenth century plant and animal geography had not made any substantial progress in finding a common law of distribution. The efforts of taxonomists, proto-ecologists and geographers only ended up creating variations of plant and animal geography. The separate origins of plant geographers especially highlight the different attitudes towards practicing plant geography from two different fields and two very different founders.

# References

Adanson, M. (1757). *Histoire naturelle du Sénégal: coquillages: avec la relation abrégée d'un voyage fait en ce pays, pendant les années 1749, 50, 51, 52 & 53*. Paris: Bauche.

Adanson, M. (1763). *Familles des plantes*. Paris: Chez Vincent, Imprimeur-Librarie de Mgr le Comte de Provence, rue S. Servin.

Allen, J. A. (1902). Zimmermann's 'zoologiae geographicae' and 'geographische geschichte' considered in their relation to mammalian nomenclature. *Bulletin American Museum of Natural History, 16*, 13–22.

Anonymous. (1789). XXVII Zimmermann's geographical history of man. *The Monthly Review, 80*, 678–690.

Baker, A. (2007). Occam's Razor in science: A case study from biogeography. *Biology and Philosophy, 22*, 193–215.

Beck, H. (1957). Geographie un Reisen in 19. Jahrhundert: Prolegomena zu einer allgemeinen Geschichte der Reisen. *Petermanns Geographische Mitteilungen, 101*, 1–14.

Beck, H. (1983). Geography and travel in the 19th century: Prolegomena to a general history of travel. In G. S. Dunbar (Ed.), *The history of geography: Translations of some French and German essays* (pp. 73–102). Malibu: Undena Publications.

Blumenbach, J. F. (1775). *De generis humani varietate nativa*. Göttingen: Rosenbusch.

Bodenheimer, F. S. (1955). Zimmermann's specimen zoologiae geographiae quadrupedum: A remarkable zoogeographical publication of the end of the 18th century. *Archives Internationales d'Histoire des Sciences, 34*, 351–357.

Bradley, M. J. (2006). *The foundations of mathematics: 1800 to 1900*. New York: Chelsea House Publishing.

Brisson, M. J. (1756). *Le regne animal divisé en IX classes*. Paris: Bauche.

Broberg, G. (1990). The broken circle. In T. Frängsmyr, J. L. Heilbron, & R. E. Rider (Eds.), *The quantifying spirit in the eighteenth century* (pp. 49–50). Berkeley: University of California Press.

Browne, J. (1983). *The secular ark: Studies in the history of biogeography*. New Haven: Yale University Press.

Buachen, P. (1757). Le parallèle des fleuves des quatre parties du monde pour servir a déterminer la hauteur des montagnes. *Histoire de l'Académie royale des sciences, 1757*, 586–588.

Cuvier, G. (1807). Sur les differentes especes de crocodiles vivans et sur leurs caracteres distinctifs. *Annales du Museum d'histoire naturelle, Paris, 10*, 8–86.

Dana, J. D. (1853). *On the classification and geographical distribution of Crustacea*. Philadelphia: C. Sherman.

Debarbieux, B. (2008). The Mountains between corporal experience and pure rationality: The contradictory theories of Philippe Buache and Alexander von Humboldt. In D. Cosgrove & V. Della Dora (Eds.), *High places* (pp. 87–104). London: IB Tauris.

de Bougainville, L. A. (1771). *Voyage autour du Monde, par la frégate du roi La Boudeuse, et la flûte l'Etoile; en 1766, 1767, 1768 & 1769*. Paris: Chez Saillant et Nyon.

de Buffon, G. L. L. C. (1749–1789). *Histoire naturelle: générale et particulière, servant de suite á l'histoire des animaux quadrupèdes*. Paris: L'Imprimerie Royale.

de Buffon, G. L. L. C. (1761). *Histoire naturelle: générale et particulière, servant de suite á l'histoire des animaux quadrupèdes* (Vol. 9). Paris: L'Imprimerie Royale.

de Buffon, G. L. L. C. (1807). *Barr's Buffon. Buffon's natural history VII*. London: Gillet.

de Candolle, A. P. (1805). Explication de la carte Botanique de la France. In: J. B. P. A. de M. Lamarck & A. P. de Candolle (Eds.), *Flore française, ou descriptions succinctes de toutes les plantes qui croissent naturellement en France, disposées selon une nouvelle méthode d'analyse, et précédées par un exposé des principes élémentaires de la botanique* (3rd ed.). Paris: Desray.

de Candolle, A. P. (1820). *Essai élémentaire de géographie botanique. Dictionnaire des Sciences Naturelles* (Vol. 18). Paris: F. Levrault.

de Candolle, A. L. P. P. (1855). *Géographie botanique raisonnée*. Paris: Masson.

# References

Erxleben, J. C. P. (1777). *Anfangsgründe der Naturlehre and Systema regni animalis*. Göttingen: Johann Hans Dieterich.
Feuerstein-Herz, P. (2004). *Eberhard August Wilhelm von Zimmermann (1743–1815) und die Tiergeographie* (pp. 1–389). Ph.D thesis, Technischen Universität Carolo-Wilhelmina zu Braunschweig.
Fichman, M. (1977). Wallace: Zoogeography and the problem of land bridges. *Journal of the History of Biology, 10*, 45–63.
Forbes, E. (1846). On the connexion between the distribution of the existing fauna and flora of the British Isles and the geological changes which have affected their area, especially during the epoch of the Northern Drift. *Memoirs of the Geological Survey of Great Britain, 1*, 336–432.
Forbes, E. (1856). Map of the distribution of marine life, illustrated chiefly by fishes, molluscs and radiata; showing also the extent and limits of the homoiozoic belts. In A. K. Johnston (Ed.), *The physical atlas of natural phenomena (Plate 31)*. Edinburgh: William Blackwood and Sons.
Forster, J. R. (1778). *Observations made during a voyage round the world, on physical geography, natural history, and ethic philosophy*. London: G. Robinson.
Geoffroy Saint-Hilaire, E. (1803a). XXIII. Anatomical observations on the crocodile of the Nile. *Philosophical Magazine Series, 1, 16*(62), 126–146.
Geoffroy Saint-Hilaire, E. (1803b). XL. Account of a new kind of American crocodile. *Philosophical Magazine Series, 1, 16*(63), 233–235.
Gill, T. (1885). The principles of zoogeography. *Proceedings of the Biological Society of Washington, 2*, 1–39.
Gloger, C. W. L. (1833). *Das Abändern der Vögel durch Einfluss des Klimas*. Breslau: August Schulz.
Gmelin, J. G. (1747). *Flora Sibirica sive Historia Plantarum Sibiriae*. Petropoli: Ex Typographia Academiae scientiarum.
Hagen, J. (1986). Ecologists and taxonomists: Divergent traditions in twentieth-century plant geography. *Journal of the History of Biology, 19*, 197–214.
Halle, J. S. (1757). *Die Naturgeschichte der Thiere in sistematischer Ordnung*. Berlin: Christian Friedrich Voss.
Hooykaas, R. (1956). The zoogeography of Abraham van der Mijle. *Archives Internationales d'Histoire des Sciences, 9*, 125–132.
Hull, D. L. (1988). *Science as process: An evolutionary account of the social and conceptual development of science*. Chicago: University of Chicago Press.
Islenev, I., & Chudjakov, E. M. (1777). *Mappa fluvii Irtisz partem meridionalem gubernii Sibiriensis perfluentis cum pristino territorio stirpis Kalmukorum songaricæ* (Russia asiatica II; Falz 23), [S.l.] [s.n.].
Jackson, S. T. (2009). Introduction: Humboldt, ecology, and the Cosmos. In A. von Humboldt & A. Bonpland (Eds.), *Essay on the geography of plants* (edited by S. T. Jackson) (pp. 1–52). Chicago: University of Chicago Press.
Jussieu, A. (1789). *Genera plantarum Secundum Ordines Naturales Disposita, Juxta Methodum in Horto Regio Parisiensi Exaratam*. Paris: Barrois.
Kames, H. H. (1774). *Sketches of the history of man*. Edinburgh: W. Creech.
Kant, I. (1775). Von den verschiedenen Racen der Menschen. In *Immanuel Kants frühere noch nicht gesammelte Kleine Schriften* (1795) (pp. 97–106). Linz: Auf kosten des Herausgebers.
Kircher, A. (1675). *Arca Noë*. Amsterdam: Johannes Jansson.
Klein, J. T. (1751). *Quadrupedum dispositio brevisqus historia naturalis*. Leipzig: Ionam Schmidt.
Kolbe, P. (1719). *Caput Bonae Spei Hodiernum*. Nuremberg: Peter Conrad Monath.
Larson, J. L. (1994). *Interpreting nature: The science of living form from Linnaeus to Kant*. Baltimore: Johns Hopkins University Press.
Linnaeus, C. (1737). *Flora Lapponicum, exhibens plantas per Lapponiuan crescentes*. London: James Edward Smith.
Linnaeus, C. (1753). *Species plantarum, exhibentes plantas rite cognitas, ad genera relatas, cum differentiis specificis, nominibus trivialibus, synonymis selectis, locis natalibus, secundum systema sexuale digestas*. Holmiae: Impensis Laurentii Salvii. L. Salvius.

Lyell, C. (1842). *Principles of geology or the modern changes of the earth and its inhabitants, considered as illustrative of geology by Charles Lyell* (Vol. 3). Boston: Hilliard, Gray & Co.
Macleay, W. S. (1819). *Horae entomologicae: Or essays on the annulose animals* (Vol. 1, Part 1). London: S. Bagster.
Maroske, S. (2012). Australian and Indian plants: Making connexions in nineteenth-century botany. *Historical Records of Australian Science, 23*, 107–119.
Martini, F. H. W. (1774). *Allgemeine Geschichte der Natur in alphabetischer Ordnung*. Berlin: J. Pauli.
Merriam, C. H. (1894). [1895] The geographic distribution of animals and plants in North America. *Yearbook of the United States Department of Agriculture, 1894*, 203–214.
Meyen, F. J. F. (1846). *Outlines of the geography of plants: With particular enquiries concerning the native country, the culture, and the uses of the principal cultivated plants on which the prosperity of nations is based*. London: Ray Society.
Minding, J. (1829). *Ueber die geographische Vertheilung der Säugethiere*. Berlin: Enslin'sche Buchhandlung.
Miracle, M. E. G. (2008). The significance of Temminck's work on biogeography: Early nineteenth century natural history in Leiden, the Netherlands. *Journal of the History of Biology, 41*, 677–716.
Mylius, A. (1667). *De Origine animalium, et migratione populorum, scriptum Abrahami Milii. Ubi inquiritur, quomodo quaque via homines caeteraque animalia terrestria provenerint; & post deluvium in omnes Orbis terrarum partes & regiones: Asiam, Europam, Africam, utramque Americam, & Terram Australem, sive Magellanicam, pervenerint*. Amsterdam: Apud Petrum Columesium.
Nelson, G. (1978). From Candolle to Croizat: Comments on the history of biogeography. *Journal of the History of Biology, 11*, 269–305.
Pallas, P. S. (1784–1788). *Flora rossica*. Petropoli: J. J. Weitbrecht.
Pennant, T. (1783). *History of quadrupeds* (3rd ed.). London: Benjamin White.
Prichard, J. C. (1826). *Researches into the physical history of mankind* (2nd ed.). London: Houlfton and Stoneman.
Ray, J. (1686). *Historia plantarum generalis*. London: Samuel Smith & Benjamin Walford.
Ray, J. (1710) Historia insectorum. *Opus posthumum Jussu Regiæ Societatis Londinensis Editum. Cui subjungitur appendix de scarabæis Britannicis, autore M. Lister S. R. S. ex Mss. Musæi Ashmolæani*. Londini: Impensis A. & J Churchill.
Roger, J. (1997). *Buffon: A life in natural history*. Ithaca: Cornell University Press.
Schmithüsen, J. (1985). Vor- un Frühgeschichte der Biogeographie. *Biogeographica, 20*, 1–166.
Schouw, J. F. (1823). *Grundzüge einer allgemeinen Pflanzengeographie*. Berlin: Reimer.
Schreber, J. C. D. (1774). *Die Säugthiere in Abbildungen nach der Natur mit Beschreibungen*. Erlangen: Wolfgang Walther.
Stromeyer, F. (1800). *Commentatio inauguralis sistens historiae vegetabilium geographicae specimen*. Göttingen: Heinrich Dieterich.
Swainson, W. (1835). *A treatise on the geography and classification of animals*. London: Longman, Brown, Green, and Longmans.
Turpin, F. H. (1771). *Histoire naturelle et civile du royaume de Siam*. Paris: Chez Costard.
von Goethe, J. W. (1813). Höhen der alten und neuen Welt bildlich verglichen. *Allgemeinen Geographischen Ephemeriden, 41*, 3–8.
von Hofsten, N. G. E. (1916). Zur älteren Geschichte des Diskontinuitätsproblems in der Biogeographie. *Zoologische Annalen Zeitschrift für Geschichte der Zoologie, 7*, 197–353.
von Humboldt, A. (1793). *Florae Fribergensis specimen*. Berlin: H. A. Rottmann.
von Humboldt, A., & Bonpland, A. (1807). *Voyage de Humboldt et Bonpland. Première partie. Physique Générale, et relation historique du voyage. Premier Volume, Contenant Essai sur la Géographie des plantes, accompagné d'un Tableau physique des régions équinoxiales, et servant d'introduction à l'Ouvrage. Chez Fr.* Paris: Schœll.

# References

von Humboldt, A., & Bonpland, A. (1829). *Personal narrative of travels to the equinoctial regions of the new continent, during the years 1799–1804* (Vol. 7). London: Longman, Rees, Orme, Brown, and Green.

von Humboldt, A., & Bonpland, A. (2009). *Essay on the geography of plants*. Chicago: University of Chicago Press.

Willdenow, C. L. (1792). *Grundriss der Kräuterkunde*. Berlin: Haude and Spener.

Willughby, F. (1686). *De Historia Piscium Libri Quatuor* (J. Ray, Ed.). Oxford: Sheldonian Theatre.

Zimmermann, E. A. W. (1777). *Specimen zoologiae geographicae, Quadrupedum domicilia et migrationes sistens*. Leiden: Theodorum Haak.

Zimmermann, E. A. W. (1778–1783). *Geographische geschichte des menschen, und der allgemein verbreiteten vierfüssigen thiere* (Vol. 3). Leipzig: Weygandschen Buchhandlung.

# Chapter 3
# Humboldt, Stromeyer and Candolle

In his paper "The Nature of Plant Geography", Hugh M. Raup (1942) argued that plant geography during the nineteenth century,

> ... has never hoped for more than an approximate solution to the problem of ultimate causation. The same would have to be said of other natural sciences such as geology and meteorology, and most of the broad field of plant and animal morphology. To hold that the logical methods of these services were "defeatist" would be denying the quality of logic that gave us the Renaissance and the development of nearly all of modern science (Raup 1942, p. 350).

Putting 'the Renaissance and the development of nearly all of modern science' to one side, Raup has a good point. The defeatist view, which he ascribes to Hartshorne (1939), seeks a universal law or process that would unify, rather than divide, the fledging field of geography. For Hartshorne, this division is the complex inter-relationships between organisms versus simple historical accounts in which these relationships remain the same.

> Even if one knew all the principles and had all the data, the solution would be involved in a mathematical equation so complicated that no finite mind could solve it (Hartshorne 1939, p. 385).

Raup poignantly outlines Hartstone's dilemma:

> In short, what prospect is there of reaching a resolution of the complex interplay of influences in a geographic area? (Raup 1942, p. 349).

Indeed, what prospect is there of any resolution when the amount of data is seemingly infinite? Naturalists started to address this problem indirectly during its classical period,[1] particularly in the works of Alexander von Humboldt, Friedrich Stromeyer and Augustin Pyramus de Candolle.

---

[1] A term Richard Hartshorne uses to describe the period of geographical thought between 1800 and 1850, the period under study in this paper: "Although the roots of geography, as a field of study, reach back to Classical Antiquity its establishment as a modern science was essentially the work

## Friedrich Stromeyer

> *The geographical history of plants, as you know, is not ignored, for it is my hobby [ ... ]
> I have a greater inclination toward chemistry and mineralogy and perhaps more talent
> (Stromeyer to his father Ernst Johann Friedrich Stromeyer, 27 September, 1801, in Beer
> 1999, p. 36).*

Nineteenth century plant geography has much to owe to two seminal works: Carl Ludwig Willdenow's *Grundriss der Kräuterkunde* (Willdenow 1792) and Alexander von Humboldt and Aimé Bonpland's Essai *sur la géographie des plantes* (Humboldt and Bonpland 1807).[2] Both works highlighted two diverging approaches to studying plants: the first was a taxonomic treatment in which all aspects of plants including their physiology, taxonomy, nomenclature and geography were listed. A reader of Willdenow's treatment had a complete anthology of plants, which was used as a reference for naturalists and collectors of plants in herbaria or areas of scholarly study. Humboldt and Bonpland's *Essai* was an attempt to break from this traditionalist approach, which had its origins in the herbal treatments of Otto Brunfels and Conrad Gessner, by quantifying plant geography. Rather than listing the places where plants are found, Humboldt hoped to show, through careful measurement of rainfall, altitudes and so on, what drives plant distribution?[3] Humboldt's approach involved going into the field, while Willdenow's approach offered an exhaustive treatment of the labours of others.

A similar practise existed in Buffon's *Histoire naturelle: générale et particulière, servant de suite á l'Histoire des animaux quadrupèdes* (Buffon 1749–1789). It too was an exhaustive treatment, distinguishing the morphological and behavioural traits of animals and where they occur in Old and New Worlds. Eberhard August Wilhelm von Zimmermann's *Specimen Zoologiae Geographicae Quadrupedum* (Zimmermann 1777), however, departed from this approach. Rather than provide a

---

of the century from 1750 to 1850. The second half of this period, the time of Humboldt and Ritter, is commonly spoken of as the 'classical period' of geography" (Hartshorne 1939, p. 221).

[2]The title page of the *Essai* is dated 1805, however there has been some debate as to the actual date of publication. Stearn (1960, p. 354) cites Humboldt's personal narrative (1: xxiv) for a 1806 publication date (see also Detwyler 1969, p. 113, Castrillon 1992, Drouin 1998), while Jackson (2009, pp. 16–17, footnote 26) states that no copies were in circulation before 1807. Stearn believes that "the difficulty of preparing and engraving the huge detailed chart issued at the same time delayed publication until 1807" (Stearn 1960, p. 354). Detwyler (1969, p. 113) also states that Humboldt was most likely the sole author of the *Essai*. The *Essai* will herein be cited as Humboldt and Bonpland (1807) in order to give some coherence to the chronological order of publications. For example, de Candolle (1805) appeared, and was most likely read, before Humboldt and Bonpland (1807), even if the imprint is dated "1805" in the latter.

[3]Much of this stems from Humboldt's earlier work, *Florae Fribergensis Specimen* (Humboldt 1793), which is predominantly a checklist of cryptogamic plants found in mines surrounding Freiburg (e.g., lichens, fungi, algae etc.). The work also includes in a section titled *Aphorisms on the Chemical Physiology of Plants*, "which contain his experiments on the susceptibility of plants, their mode of nourishment their colour, etc." (Bauer 1852, p. 21; Bowen 1981, p. 214).

general handbook, it only focused on the distribution of quadrupeds based on what was understood at the time. Zimmermann provided the first anthology of animal distribution.

Published in 1800, *Commentatio Inauguralis Sistens Historiae Vegetablium Geographiae Specimen* (herein *Specimen*)[4] by Fredrich Stromeyer (1776–1835) is the first and most complete anthology of eighteenth century plant geography, surpassing Willdenow's *Grundriss der Kräuterkunde*. Stromeyer's *Specimen*, which closely followed Zimmermann's *Specimen Zoologiae Geographicae Quadrupedum*, was also the first doctoral dissertation on plant geography, which contained the beginnings of a larger work, which was heralded as:

> ... grasp[ing] the subject to the fullest extent, and has opened new visionary insights [...] His trail-blazing research is, in many respects, of the upmost interest (Schrader 1800, p. 385; also in Beer 1999, p. 9).

> ... the beginning of a larger work, from which the author provides an overview of the topic, although a start, is entitled to greater aspirations (Gmelin 1801, p. 1696; also in Beer 1999, p. 8).

Regardless, Stromeyer produced no further work on plant geography, instead turning to chemistry and mineralogy, to which he would later significantly contribute (see Beer 1999, Fig. 3.1). While it is surprising that a doctorate in plant geography led to a career in mineralogy, it is the reaction to his *Specimen* several years later, which is of greater interest, perhaps most notably for souring the relationship between von Humboldt and Augustin Pyramus de Candolle. Given Stromeyer's impact on early nineteenth century plant geography and chemistry, there is a small but informative body of literature about his life and achievements in chemistry.[5]

## The Scope of Stromeyer's *Specimen*

The initial reaction to Stromeyer's thesis was minimal. In 1800, Heinrich Adolf Schrader, a botanist and editor of the *Journal für die Botanik*, praised Stromeyer's *Specimen* in a glowing nine page review, while a year later Johann Friedrich Gmelin managed a modest half page review in the *Göttingischen Anzeigen von gelehrten Sachen*. At the time of publication of Stromeyer's *Specimen*, Humboldt was on his famous South American journey, which was the inspiration for the subsequent publication of his *Essai* (co-authored with Aimé Bonpland) in 1807 and

---

[4] Stromeyer referred to his thesis as his "*Specimen*" (see Beer 1999).

[5] For a detailed genealogy of Stromeyer see von Wilcke (1967). Various biographies of his career are covered by Beer (1999, 2006), Lockemann and Oesper (1953), Thomson (1830), and shorter pieces by Asimov (1972).

**Fig. 3.1** *Carte Botanique de France* of Lamarck and de Candolle (1805, third edition, vol. 2). The map reads *Carte Botanique de France pour la 3ème Edition de la Flore française par A.G. Dezauche fils Ingénieur Hydrogéologue de la Marine an 13* (1805) (Image courtesy of Erin Clements Rushing and the Smithsonian Institution Libraries)

later works.[6] Humboldt's Essai however, had completely overlooked Stromeyer's contribution. For Humboldt, his *Florae Fribergensis Specimen* of 1793 was the last word on plant geography,[7]

---

[6] Humboldt's later works focus largely on observations and measurements made during his American journey (1799–1804). For a full list see Jackson in Humboldt and Bonpland (2009, pp. 262–263).

[7] The text referring to plant geography appeared in the introduction and as an explanatory footnote given above (Humboldt, 1793, ix–x). In the preface to the *Essai* Humboldt mentions giving "a first sketch of a geography of plants in 1790 to Cook's famous colleague, Mr. Georges [sic] Forster, with whom I had close ties of friendship and gratefulness" (Humboldt and Bonpland 2009, p. 61; original text in French Humboldt and Bonpland 1807, p. vi)". Could this sketch be the same footnote that appeared in Humboldt's *Florae Fribergensis Specimen* of 1793? Certainly by 1790 Humboldt was pondering the laws of plant distribution: "Not every rock is a habitat for every plant. Nature follows unknown laws which can only be explored further once botanists submit more data for inductive reasoning" (Humboldt 1790, p. 86).

> Geonosy (*Erdkunde*) studies animate and inanimate nature [...] both organic and inorganic bodies. It is divided into three parts; solid rock geography, which Werner has industriously studied: zoological geography, whose foundations have been laid by Zimmermann; and the geography of plants, which our colleagues have left untouched. Observations of individual parts of trees or grass is by no means to be considered plant geography; rather plant geography traces the connections and relations by which all plants are bound together among themselves, designates in what lands they are found, in what atmospheric conditions they live, and tells of the destruction of rocks and stones by what primitive forms of the most powerful algae by what roots of trees, and describes the surface of the earth in which humus is prepared. This is what distinguishes geography from nature study, falsely called nature history; zoology (*zoognosia*), botany (*phytognosia*) and geology (*oryctognosia*) all form parts of the study of nature, but they study only the forms, anatomy, processes, etc., of individual animals, plants, metallic things or fossils. Earth history, more closely affiliated with geography than with nature study, but as yet not attempted by any, studies the kinds of plants and animals that inhabited the primeval earth, their migrations and disappearance of most of them, the genesis of mountains, valleys, rock formations and ore veins the earth surface gradually covered with humus and plants, denuded again by violent stream floods, and once more dried and covered by grass. Thus zoological history, the history of plants, and the history of rocks, which tell only the past state of the earth, are to be clearly distinguished from geography (Humboldt in Hartshorne 1958, p. 100, original emphasis; original in Latin Humboldt 1793, pp. ix–x, footnote *).

However, unbeknownst to Humboldt, Stromeyer had not only addressed plant geography but had created a classification of the field and a lengthy anthology of its practitioners:

> There is practically no other part of Botany treated with less zeal and more neglected and less esteemed than the Geographic History of Vegetables, and from that fact in no part of this science is our knowledge found to be so defective and imperfect as in this; inasmuch as up to now it has arrived at a very poor stage of elaboration. The anatomy and physiology of plants, though from the evil character of the time little cultivated and nearly always little cared for, have had nevertheless certain cultivators and promoters here and there. This truly has scarcely happened to the Geographic History of Vegetables, though it is a subject most worthy to be known and understood. For if we survey and fly through the long series of botanical writings, from the origin of the science up to the times of the great Hedwig and his contemporaries, we will light upon not one name which in any way can rival or follow Zimmermann (Stromeyer 1800, pp. 15–16).

Compare this in strong contrast to Humboldt's own views on plant geography in his 1807 *Essai*:

> ... it is no less important to understand the Geography of plants, a science that up to now exists in name only, and yet is an essential part of general physics (Humboldt and Bonpland 2009, p. 64).

Humboldt clearly viewed himself as the founder and first practitioner of plant geography,[8] seeing no other attempts before his own. This can partly be explained by his hands-on approach, as Humboldt was in a position to practise plant geography.

---

[8] While Humboldt did not directly title himself the as the founder, he did elude that he was the first practitioner of plant geography "... a science that until now exists in name only ..." (Humboldt and Bonpland 2009, p. 64).

As a wealthy aristocrat with the means to travel, Humboldt put his ideas into practise at great expense, spending a third of his inheritance traveling and collecting measurements and creating numerous maps (de Terra 1955, p. 198, footnote 1). Far beyond the opulence of Humboldt's world, Stromeyer had little option but to study plant geography theoretically, looking at its goals, its practitioners and compiling a vast literature review. Given the different attributes of each study, Humboldt's empirical *Essai* cannot be directly compared to Stromeyer's theoretical *Specimen*. For instance, Humboldt had climbed the Chimborazo, one of the highest peaks in Latin America and which was later to feature as the backdrop to his famous *Tableau physique*.[9] There he collected specimens and measured elevation, temperature and the organisms that reside there, while Stromeyer simply had access to a large volume of literature, which he divided into:

I. What properly are to be regarded as sources of the Geographic History of Vegetables, and particularly,
   A. Specimens of the Geographic History of Vegetables already published.
   B. Commentaries and sources concerning single heads and parts of the Geographic History of Vegetables.
   C. Sources and writings in which is treated the Geography of Vegetables of single regions of the earth; to which pertain:
      a. Botanical topographies or Floras.
      b. Physical topographies.
      c. Descriptions of travels.
   D. Writings comprising the Geography of single Vegetables (Families, Genera and Species).
II. Aids to the Geographic History of Vegetables. Here are referred writings which treat Geography, natural History and the History of man (Stromeyer 1800, p. 20).

For Humboldt this may not have been a great empirical or analytical feat, however Stromeyer did encapsulate the Geographic History of Vegetables into a synthesis, namely:

The discussion of this matter rests upon these questions:

1. How the incredible multitude of Vegetables and their forms and conformations, endlessly various and multiform, spread and are distributed over the earth at the present day; what are the laws of their spread and distribution. – Geography of Vegetables, Phytogeography.

---

[9]The *Tableau physique* is a large plate featured in Humboldt and Bonpland's *Essai*. "With this plate, however, Humboldt invents a complex mode of visualizing scientific data, corresponding to his new science of the geography of plants, itself a part of his idea of a future of general physics" (Romanowski, in Humboldt and Bonpland 2009, p. 157).

2. Whether Vegetables formerly occupied the earth in the same way as at the present day, or if that is truly the case, and what they may have undergone before they arrived at that station in which they are now found; what causes provide a place for these changes, and what things follow from this. – Geographic History of Vegetables.
3. Lastly, by what relation all these things agree with the history of the earth and the rest of its inhabitants, both man and animals. – Applied Geographic History of Vegetables.

After the example of the celebrated Zimmermann [1777, 1778–1783], who so excellently published on a similar subject in the matter of animals, I believe that I can gather the whole outline of this subject, so full and abundant, scarcely inconveniently under this common notion, and encompass it by the denomination of the Geographic History of Vegetables (Stromeyer 1800, pp. 14–15).

Stromeyer's classification of historical plant geography was at odds with Humboldt's plant geography,

Observations of individual parts of trees or grass is by no means to be considered plant geography; rather plant geography traces the connections and relations by which all plants are bound together among themselves, designates in what lands they are found, in what atmospheric conditions they live, and tells of the destruction of rocks and stones by what primitive forms of the most powerful algae by what roots of trees, and describes the surface of the earth in which humus is prepared (Humboldt in Hartshorne 1958, p. 100).

Stromeyer divided plant geography into categories based on what people had written previously, thereby creating an anthology. In contrast, Humboldt created a guide for implementing his plant geography that sought to look at a vast array of different interconnections through a classification of vegetation types. In doing so, Humboldt created a cosmology, which generally lacked a clear theory with succinct aims or goals.[10] Not until Joakim Frederik Schouw's *Grundzüge einer allgemeinen Pflanzengeographie* (1823), did Humboldt's work attain some degree of significance in terms of theory. Schouw, like the 'Humboldtians' that followed him, created new methodologies and new fields of inquiry, such as ecology (see Nicolson 1996). Moreover, by writing his thesis in Latin, Stromeyer had pipped Humboldt by unwittingly coining the term phytogeography.[11] However, it was not until 1814 that Humboldt had finally heard of Stromeyer's *Specimen*.[12] Humboldt's reaction may have soured his relationship with de Candolle.

---

[10]Humboldt's later work also lacked a clear definition, "Humboldt has, in outlining the factors belonging to plant geography in his *Essai* and *Prolegomena*, probably not have had the discretion to give a definition" (Schouw 1823, p. 7, my emphasis).

[11]Humboldt also wrote his *Florae Fribergensis* in Latin, but had referred to the geography of plants as *Geographia planatarum* (Humboldt 1793, p. ix, and in footnote).

[12]Given the geographical distance between nineteenth century researchers, it took time for manuscripts and folios to circulate. Humboldt did well to publish in French and spend a majority of his time in Paris (see Stearn 1960, p. 353), a point that Goethe concedes to Eckermann on May 3, 1827: "it has not been so easy; and we others also, in Central Germany, have been forced to

## Candolle's Grudge

Augustin Pyramus de Candolle's relationship with Alexander von Humboldt has always been considered to be informal and friendly (see Drouin 1998, 2004). They exchanged a few letters and mostly spoke during meetings at the influential *Société d'Arcueil* – "a scientific association, composed of some of the most distinguished savans of Paris" (Stoddard 1859, p. 345):

> Gay-Lussac, Thenard, Decandolle, Collet, Descotils, Malus; A. B. Berthollet, and Humboldt [ . . . ] met once a fortnight at the house of Berthollet, and spent the day together, giving each other the results of their studies and experiments, reading the scientific papers that they had composed since their last meeting, or in pleasant rambles about the neighbourhood. Most of these men were members of the Institute of France, and the papers that they read at Arcueil, were delivered before that august body, and afterwards published in the 'Memoirs' of the society (Stoddard 1859, p. 346).

In his *Mémoire sur la géographie des plantes de la France*, A.P. de Candolle (1817) praised Humboldt and his new science:

> The geography of plants is almost a new science; although it had been attended to by Linnaeus and some of his successors, yet the first writer, who can be considered as having treated upon it in a regular and systematic manner, is M. De Humboldt (Candolle 1817, p. 408).

Yet it comes as a surprise, when reading A. Candolle's (1862) autobiography that he had other feelings for the German aristocrat:

> Humboldt also came from time to time; but he added much of life and interest when he appeared [possibly 1806 or 1807]. He affected to pass himself as the creator of the science of botanical geography, to which he has only added certain facts, and the exaggeration of a true theory so as to render it almost false. He never quite pardoned me for having, in the preface to my memoir on the geography of the plants of France [ . . . ], cited those who before him had occupied themselves with geographical history, although in his exposition

---

buy our little wisdom dearly enough. Then we all lead a very isolated miserable sort of life! From the people, properly so called, we derive very little culture. Our talents and men of brains are scattered over the whole of Germany. One is in Vienna, another in Berlin, another in Königsberg, another in Bonn or Düsseldorf – all about a hundred miles apart from one another, so that personal contact and personal exchange of thought may be considered as rarities. I feel what this must be, when such men as Alexander von Humboldt come here, and in one single day lead me nearer to what I am seeking and what I require to know than I should have done for years in my own solitary way.

But now conceive a city like Paris, where the highest talents of a great kingdom are all assembled in a single spot, and by daily intercourse, strife, and emulation, mutually instruct and advance each other; where the best works, both of nature and art, from all the kingdoms of the earth, are open to daily inspection; conceive this metropolis of the world, I say, where every walk over a bridge or across a square recalls some mighty past, and where some historical event is connected with every corner of a street" (Goethe 1883, p. 252).

I had, in truth, much amplified his share (Candolle in Gray 1889, p. 303; from Candolle 1862, p. 167; see Drouin 1998, p. 11 for alternate translation).[13]

Perhaps Candolle's grudge was disguised during those meetings, only to erupt and appear in print a decade after Humboldt's death, during the height of the ever-popular *Kosmos*. One could imagine that the meeting had a Pythonesque feeling to it – an outspoken iconic hero of the day, outwardly admired by his less travelled colleagues who, at the same time, were repressing an apparent inward disdain. Candolle's grudge however is understandable. In 1805, a year before *Arcueil*, Candolle was 27 and the son of a provincial Swiss banker. Humboldt was 36, an aristocrat, trained in economics and administration and was world famous for having described the natural history of the New World. Candolle may have interpreted Humboldt's popularity, flair, ambition and modesty as dismissing lesser-known natural historians who for the most part worked on known or smaller problems, but who also had discussed plant geography.

The communication between Humboldt and de Candolle during the early part of the nineteenth century was cordial even to the point that Humboldt requested Candolle to remove a line from his forthcoming *Memoire sur la Geographie de plantes de France*, recently read at the *Arcueil*, which had introduced Stromeyer's *Historiae Vegetablium Geographiae Specimen* as a primary work prior to Humboldt's 1807 *Essai*:

> Your entire memoir shows great kindness to me; I am mentioned on every page, but there is mention of diversity of opinions – words kind and friendly as is your character – that has no place in a volume published by men so intimately related. But this history: I suffer but not alone because it leaves the same impression on many of our friends. This is my hope, my dear Candolle, see if it agrees with your ideas about historical truth: look again at Strohmayer, to see if his book contains a *number – a measure –* to see if it is any more than a mass of citations. I was not the first to envision the geography of plants. When men first talked of "alpine plants," they framed the basis of that science. You need not change what you have written about Giraud Soulavie, who already has written about the altitudinal limits

---

[13]The apparent grudge lasted until Candolle's *Memories*, as his son Alphonse contributes his own commentary to the above passage in two footnotes (below and in main text):

> The author no doubt intended to write about the influence of the elevation of the soil on the distribution of plants; but the word theory is not appropriate, because the defect in the works of Humboldt, in botanical geography, is precisely the absence not only of theory, but even of any discussion of the facts. The most original thing in his works is the isothermal lines, of which he originated the idea and highly developed it; but these lines of equal temperature belong to the science called terrestrial physics or geographical physics. I have shown that they are entirely without application to the phenomena of geographical botany and of agriculture, for which maxima, not averages, explain the facts. (de Candolle 1862, p. 167, footnote 1 by Alphonse de Candolle. Translated by G. Nelson).

of olive trees and traced them for the Vivarais province; he deserves such praise. You need not change a word about M. Strohmayer [sic], who developed some of the ideas, which I announced in 1794 in the *Flora Fribergensis*, on the migrations of plants and the real difference between the *geography* of plants and their history – the summary of the events that permitted migrations. All that I ask of you reduces to the observation that I gave the first botanical map [*Tableau physique*, 1807] and the first work founded on real measures of altitude and observations of temperature (Humboldt to de Candolle dated 1814, in de la Roquette 1865, pp. 197–199, translated by G. Nelson, original emphasis).

Humboldt's request was florid and biting, mostly toward Stromeyer[14] whom Humboldt compared to de Candolle's one time critic Charles-François Brisseau de Mirbel,

Everyone has his Mirbel, M. Strohmayer [sic] is mine. You will not say, I hope, that my [request] resembles a quarrel. Quarrels, you know, give life to the Academies (Humboldt to de Candolle dated 1814, in de la Roquette 1865, p. 197, translated by G. Nelson).

Give life to the academies it did. Two years later in 1816 on February 5th, Humboldt put his objections to the French Institute in his *Sur les lois que l'on observe dans la distribution des formes végétales* or *On the Laws observed in the Distribution of vegetable Forms*,

His Geography of tile Plants in the South of France was followed by the *Tentamen Historiae geographiace Vegetabilium* of the learned Professor Strohmayer [sic], published in 1800 at Gottingen in the form of a dissertation; but this Tentamen exhibits rather the plan of a future work, and the catalogue of authors to be consulted, than information respecting the altitudes which spontaneous plants reach in different climates. The case is the same with the very philosophical views announced by M. Treviranus in his *Essai de Biologie*; we therein find general considerations, but no measurements of heights, and no thermometrical indications, which are the solid bases of the geography of plants. This study, has not risen to the rank of a science, until men of science have perfected both the measures of heights by barometrical observations, and the determination of mean temperatures; or, what is more important for the development of vegetation, the determination of the differences between the temperature of summer and winter and between that of day and night (Humboldt 1816a, p. 447 and 1816b, pp. 226–227 [in French]).

But one can almost feel the anger of de Candolle at the mention of Humboldt's map "the first work based on real measures and observations on temperature" and of plant geography, namely "this study, has not risen to the rank of a science, until men of science have perfected both the measures of heights by barometrical

---

[14]Candolle also had his detractors. French botanist, Charles-François Brisseau de Mirbel appeared to dislike Candolle's (1813) *Théorie élémentaire de la botanique*, something he told fellow botanist Horace Benedict Alfred Moquin-Tandon "… during my last trip to Paris, that the elementary theory of M. De Candolle was an absurd book" (Moquin-Tandon to de Saint-Hilaire in Leotard 1893, pp. 86–87, my translation, original emphasis; see Stevens 1994, p. 521, footnote 40.) Moquin-Tandon statement clarifies Humboldt's admission to de Candolle: "Everyone has his Mirbel, M. Strohmayer is mine". "You will not say, I hope, that my [request] resembles a quarrel" says Humboldt "Quarrels, you know, give life to the Academies".

observations, and the determination of mean temperatures". Candolle published first floristic distribution map in 1805, in the second volume of the third edition of the *Flore française* (Fig. 3.1):

> The botanical map of France [*Carte Botanique de France*], which we thought would be a useful addition to this book, is designed to highlight two very different things: (1) the knowledge of vegetation in different parts of France that are known by Botanists; and (2) the general plant distribution on French soil (Lamarck and Candolle 1805, p. i, translated in Ebach and Goujet 2006, p. 766).

The *Carte Botanique de France* is a monumental achievement, as it attempts to define the distributions of French flora based on temperature, hydrology and soil mobility. Ironic that Humboldt should ask de Candolle to bolster his claim above all others, when the Prussian aristocrat himself fails to mention the achievements of his peers.[15] De Candolle's grudge may also have been amplified by Humboldt's usage of "Strohmayer" throughout his 1814 letter. Given this obvious misspelling, one can almost be certain that Humboldt had not seen Stromeyer's *Historiae* but only heard it said, most likely in de Candolle's presentation to the *Arcueil*.[16] In a footnote in de Candolle's biography, his son Alphonse reasserts his father's claim:

> I sent M. de la Roquette many letters from Humboldt to my father, which were to be published in a book of letters written in French by this illustrious explorer. One of the longest and most interesting is a refutation of the subject of the preamble to the *Mémoire sur la Géographie* of plants of France, as it was read at Arcueil. A word that displeased Humboldt was deleted before printing. Besides, for anyone who knows the character of Humboldt and his works, the gracious modesty of his letters detracts not from what my father says about his pretensions (de Candolle 1862, p. 167, footnote 2 by Alphonse de Candolle. Translated by G. Nelson).

De Candolle did not yield to Humboldt's request rather Stromeyer's thesis was introduced prior to the German aristocrat as a "plan of a book on plant geography which is likely to announce the scope of the science" (de Candolle 1817, pp. 262–263, my translation). Moreover, in the 1817 printing of *Mémoire sur la géographie des plantes de la France*, de Candolle misspelt Humboldt throughout as "Humboltd", possibly in light of Humboldt's misspelling of Stromeyer.

De Candolle's grudge is of a sociological nature limited to a small botanical community fractured into factions. The same is true for Humboldt and his reaction to Stromeyer. Regardless of the florid, almost comical nature of his request, Humboldt clearly wanted it removed. But why? It surely was not for lack of respect. De Candolle credited Humboldt as "But the most essential work we have is that of M de Humboldt. This scientist has given on the geography of plants a work as remarkable for its great number of new facts, as by their intimate

---

[15] See Browne (2002), p. 124, footnote 60.

[16] A native German speaker would spell Stromeyer with an 'h' and an 'a', that is "Strohmayer". Given that Humboldt may not have seen Stromeyer written out, I assumed that Humboldt had not seen Stromeyer's *Specimen* before 1814. Surprisingly German geographer Gerhard Engelmann (1966) also misspells Stromeyer as "Strohmayer" (Engelmann 1966, p. 109).

connection with the most important physical laws" (de Candolle 1817, p. 263. Translated by G. Nelson). Perhaps it was a notion of one-upmanship. Humboldt most likely discovered the existence of Stromeyer's *Historiae* in 1814 at the *Arcueil*, while listening to de Candolle's *Mémoire sur la Géographie*. Not knowing about Stromeyer or the existence of his *Historiae* must have startled Humboldt.[17] A complete omission of a recent synthesis may have led to a *cause célèbre* in *Arcueil* circles. By 1814 Humboldt's *Essai* was published in at least two languages with no mention of Stromeyer. Had Humboldt known of Stromeyer's attempt at a synthesis, the beginning of Humboldt and Bonpland's *Essai* would have been dramatically different. Given the physical and cultural distances between Paris and the provincial German speaking capitals, it is no wonder that the *Historiae* remained unknown until 1814. The *Essai* was completed in 1805 and presented to the Paris *Classe des sciences physiques et mathématiques* (Jackson 2009, pp. 16–17, footnote 26), contains a smattering of literature, for example, Lamarck and Candolle's remarkable *Carte Botanique de France*, a map the precedes that of Humboldt's *Tableau physique* was not mentioned. While Stromeyer may be bane of Humboldt, for de Candolle Humboldt certainly was his new "Mirbel".

## The Monopoly of the Ages: The Rise of Humboldt and the Fall of Stromeyer

Fredrich Stromeyer (1776–1835), later famous for discovering cadmium, completed his medical degree at the University of Göttingen in 1800 on plant geography under the famous botanist Johann Friedrich Gmelin (Beer 1999; Lockemann and Oesper 1953). Stromeyer's dissertation was published as *Historiae Vegetablium Geographiae Specimen,* namely "a catalog of writings to be considered as sources of the Geographic History of Vegetables, to briefly add information about those things that are contained by them and pertain to our scope, and to say a word about their merit and worth" (Stromeyer 1800, p. 19, Fig. 3.2). Humboldt however regarded his own *Floræ Fribergensis* as the earlier and superior work that "You need not change a word for M. Strohmeyer [sic], who has developed some of the ideas that I exposed in the Flora Fribergensis on the migrations of plants, and on the difference between the geography of plants and their history – the particular events that favored the migrations" (de la Roquette 1865, p. 199. Translated by G. Nelson). Stromeyer certainly knew of Humboldt's *Floræ Fribergensis,* however, and to Humboldt's dismay, stated that "[t]here is practically no other part of Botany treated with less zeal and more neglected and less esteemed than the Geographic History of Vegetables" (Stromeyer 1800, p. 15). Stromeyer saw himself in the

---

[17]Possibly unbeknownst to Humboldt at the time, Stromeyer had coined the term 'phytogeography' (Stromeyer 1800, p. 14).

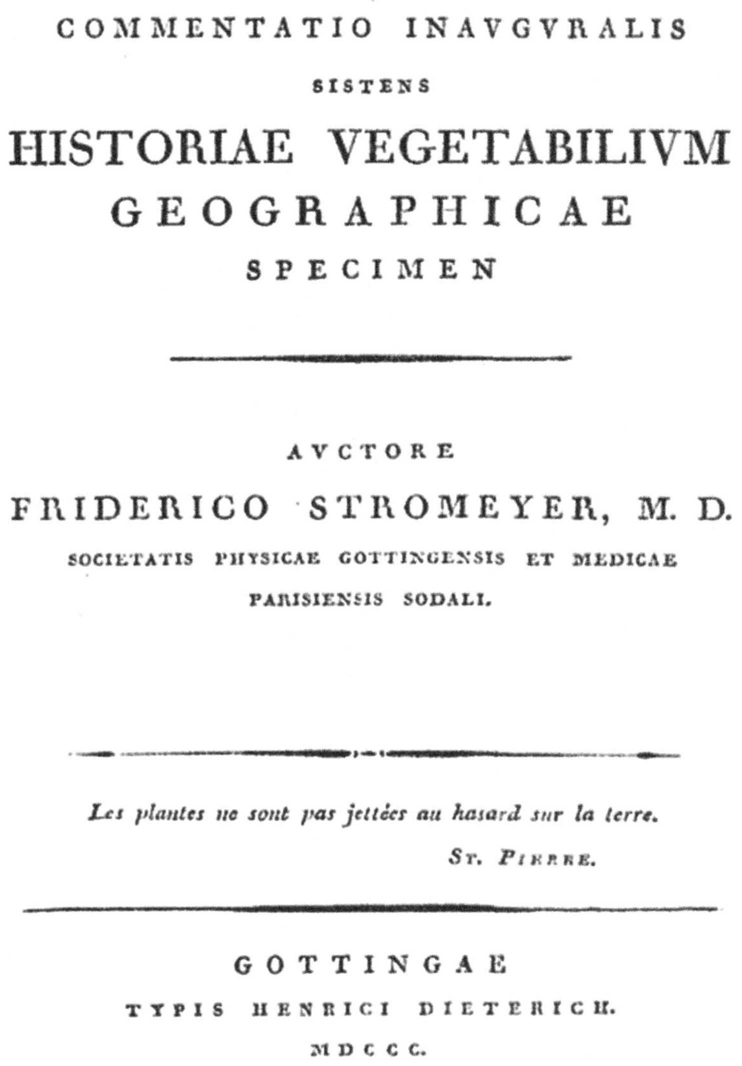

COMMENTATIO INAVGVRALIS

SISTENS

# HISTORIAE VEGETABILIVM
# GEOGRAPHICAE

SPECIMEN

AVCTORE

FRIDERICO STROMEYER, M. D.

SOCIETATIS PHYSICAE GOTTINGENSIS ET MEDICAE

PARISIENSIS SODALI.

*Les plantes ne sont pas jettées au hasard sur la terre.*

St. Pierre.

GOTTINGAE

TYPIS HENRICI DIETERICH.

MDCCC.

**Fig. 3.2** Cover of *Commentatio inauguralis, sistens historiae vegetabilium geographicae specimen* by Friedrich Stromeyer (1800). A translation of Stromeyer's introduction is in the Appendix

role of an aggregator, someone who complies all the writings on botany, classified into types of geography, in order to create a historical phytogeographical synthesis (Fig. 3.3):

**Fig. 3.3** Friedrich Stromeyer (1776–1835). Lithograph (27 × 36 cm) by Eduard Ritmüller (1805–1868), Museum der Göttinger Chemie, Fakultät für Chemie, Georg-August-Universität Göttinge (Source: Günther Beer)

> What is the Geographic History of Vegetables, and what things and ideas are comprehended under this general denomination.
> 
> The discussion of this matter rests upon these questions:
> 
> 1) How the incredible multitude of Vegetables and their forms and conformations, endlessly various and multiform, spread and are distributed over the earth at the present day; what are the laws of their spread and distribution. – Geography of Vegetables, Phytogeography.
> 2) Whether Vegetables formerly occupied the earth in the same way as at the present day, or if that is truly the case, and what they may have undergone before they arrived at that station in which they are now found; what causes provide a place for these changes, and what things follow from this. – Geographic History of Vegetables.

3) Lastly, by what relation all these things agree with the history of the earth and the rest of its inhabitants, both man and animals. – Applied Geographic History of Vegetables.

After the example of the celebrated Zimmermann, who so excellently published on a similar subject in the matter of animals, I believe that I can gather the whole outline of this subject, so full and abundant, scarcely inconveniently under this common notion, and encompass it by the denomination of the Geographic History of Vegetables. (Stromeyer 1800, pp. 14–15).

Under his title "Specimens of the Geographic History of Vegetables already published" Stromeyer lists Abbé Giraud Soulavie (1783) and Karl Ludwig Willdenow (1792), "[i]ndeed, Menzel, Adanson, Forskål and Zimmermann already took up certain thoughts on the Geography of Vegetables, which in full justice it is owed to our author to have refined and perfected and it is credited to him as first" (Stromeyer 1800, p. 22). Humboldt must have been livid. Relegated to a footnote on pages 49 and 50, Humboldt was not mentioned in the introduction along with influential names like Linnaeus, Zimmermann, Willdenow, Brown and the Forsters. Neither was there any mention of Humboldt's plant geography outlined in the *Floræ Fribergensis*. Stromeyer's supervisor, renowned botanist Johann Friedrich Gmelin (1748–1804), however, would have pointed out Humboldt's contribution. So would Heinrich Adolf Schrader, also a botanist and editor of the *Journal für die Botanik*, who in 1800 [1801] wrote a glowing nine page review of Stromeyer's thesis (Schrader 1800 [1801]). Most likely Humboldt's *Floræ Fribergensis*, like Stromeyer's *Historiae,* were over looked. Publishing his *Essai* in both French and German most likely helped proliferate Humboldt's plant geography, while Stromeyer's *Historiae* was most likely popularised by de Candolle (1817). Nevertheless, Stromeyer seemingly disappeared from the botanical literature, leaving to pursue a career in chemistry.[18]

A year later, however one of the few critiques of Humboldt's plant geography appeared as a 50 page anonymous letter[19] in the first volume of *Jahrbücher der Gewächskunde* edited by Kurt Sprengel, Scharder and Heinrich Freidrich Link in 1820 titled "Some remarks about two works concerning the Plant Geography by Mr. von Humboldt. In a letter to the Counsellor Schrader, by ..." (Anonymous 1820,

---

[18]Interestingly, Stromeyer was the first recorded doctoral student in plant geography and, had he continued on as a professor of botany, he would have been the first fully trained professional phytogeographer. While this claim may seem futile (Stromeyer went on to become a chemist), no contemporaneous professor of botany during this time, such as Göran Wahlenberg and Joakim Frederik Schouw, had a doctorate in plant geography.

[19]In the Preface, Link notes "... that it goes without saying that the editors do not always agree with the opinions expressed in some articles, especially since much of the literature is not in mutual agreement. We will print submitted articles unaltered, and without comment, however corrections of the authors will be granted. Detailed rejoinders are treasurable contributions. Incidentally publishers of such magazines have the right to refuse the printing of papers and comments which they deem useless for the purpose of the journal. Berlin, 30th March 1818 (Link 1820, pp. iii–iv)". Given that the anonymous letter is the only rejoinder within the issue, one wonders if Link has written an early disclaimer?

p. 6). In it, accusations fly with particular reference to the geography of plants as discussed in Humboldt's *Prolegomena* (Humboldt 1815) and *Observation of plant forms* (Humboldt 1816b):

> ... one cannot by any means approve of this view of the subject [Geography of Plants]; because, being merely an enumeration of the chief points which constitute the science, no advantage is gained by it. The examination of the natural affinities between plants, or, in other words, the natural arrangement of plants, belongs to the philosophy of botany [...] The author should give such a definition of the geography of plants, in the year 1793, was not very blameable considering the state of the science at that period; but that he should at this time repeat it, when Professor Stromeyer has so fully and satisfactorily established the objects of this branch of science, and when so much has been done in it by that gentleman and others, is so much the more surprising, as there is a striking difference between his own Essay and the present Treatise, in this respect (Anonymous 1821, 237–238, original emphasis).[20]

Any article referring to a largely overlooked thesis in 1820 (and again in English in 1821) is interesting, especially when the anonymous author starts critiquing the differences between the geography and the history of plants, the former being,

> ... that science which teaches us to know the APPEARANCE, DISSEMINATION, AND DISTRIBUTION OF PLANTS, AS THESE EXIST AT PRESENT WITH A DUE CONSIDERATION OF OTHER MATTERS CONNECTED WITH THEM. It considers the different *habitats* of plants, and the distinction between those kinds which are social and those which are solitary, as well as between such as are plentiful and such as are rare; which is perhaps sufficiently expressed by the word (vorkommen) occurrence. It determines the extent of districts over which plants are spread; and the laws according to which not merely the whole vegetable world, but likewise particular families and genera, are distributed in respect to geographical longitude and latitude, altitude etc. It borrows from physics and physiology the laws, according to external circumstances, as soil, temperature, moisture etc., act upon vegetables, for the purpose of comparison with those by which the geographical distribution etc., are governed [...] The HISTORY OF PLANTS, on the other hand, teaches us THE LAWS, THE VARIETIES AND THE DECAY OF THEIR ORGANIZATION. This science, also, resolves the questions, When, where and how the vegetable were first produced? To what extent are we justified in admitting the transportation of plants? Have old species disappeared, and new ones been produced? Is it possible that one species, through the influence of external causes, or through hybrid generation, can be converted into another?

---

[20] A translation of the anonymous letter was published in the *Quarterly Journal of Science, Literature and Art* in 1821. A brief introduction to the letter was given by a certain J.F.D., who like most named authors of the time praised Humboldt: "The varied and extensive information of this philosopher is well known, and justly appreciated; but the extreme vivacity and brilliancy of his imagination, and the propensity to generalize, which he manifests upon all occasions, are too conspicuous, not to excite our doubts respecting the accuracy of some of his conclusions. How far we are justified in this, the following critical and illustrative remarks will shew [sic]" (J.F.D. 1821, p. 236). J.F.D. is most likely to be chemist John Frederic Daniell (1790–1845), who was a regular contributor to the journal. The link between Daniell and Stromeyer is not clear other than they may have corresponded as chemists.

> This distinction appears to me the more natural and proper, because the geography of plants is founded wholly upon observation; whereas a part of the history of plants rests upon hypothesis. We may the certainly regard [these] as separate branches of science (Anonymous 1821, 239, original emphasis).

Stromeyer himself most likely wrote the anonymous note published by Schrader. After all, both were at University of Göttingen during the same period, and Schrader was quite favourable in his review of Stromeyer's *Historiae*, which he considered "a breakthrough" (Schrader 1800, p. 385; see also Beer 1999, p. 9). Schrader's published note of 1818 (and 1821) was the last we hear of Stromeyer.

This outburst was highly unusual. Humboldt commanded a deep respect among botanical community. But botany was not all that Humboldt dabbled in. Another anonymous review, this time of Humboldt's *Essai Géognostique sur le Gisement des roches dans les deux Hémisphères* (Humboldt 1823), had this to say:

> We have said it was a justice we owed to others, to shew [sic] that M. Humboldt was not the only authority, or the highest, in every subject of science. And it is not fitting or just, that the public, which cannot judge, and which necessarily follows the cry of the day, should measure any man by such an imaginary standard, and set up a false god to worship, to the debasement of all others. It is an age of monopolies; but it is hard that the principle of monopoly should be extended from fish to fame, from tea and porter to geognosy and botany, and to all else which ought to be the common property of the republic (which we hope it will ever continue) of science and letters (Anonymous 1825, p. 327).

The "republic of science and letters" was indeed in some strife. The aims and goals of plant geography clearly remained an issue between the classification of Humboldt and that of Stromeyer and de Candolle:

> By the term *station* I mean the special nature of the locality in which each species customarily grows; and by the term *habitation, a* general indication of the country wherein the plant is native. The term station relates essentially to climate, to the terrain of a given place; the term habitation relates to geographical, and even geological, circumstances [...] The study of stations is, so to speak, botanical topography; the study of habitations, botanical geography [...] The confusion of these two classes of ideas is one of the causes that have most retarded the science, and that have prevented it from acquiring exactitude (de Candolle 1820, p. 383, translated in Nelson 1978, p. 280).

De Candolle's definition of botanical geography is in complete opposition to Humboldt's ".... botanical geography, which assigns to each tribe of plants their height, limits, and climate". Here the *Historiae* "of the learned Professor Strohmayer [sic] [...] exhibits the plan of a future work, and the catalogue of authors to be consulted, than information respecting the altitudes which spontaneous plants reach in different climates. The case is the same with the very philosophical views of announced in M. Treviranus in his *Essai de Biologie*; we therein find general considerations, but no measurements of heights, and no thermometrical indications, which are the solid bases of the geography of plants. This study, has not risen to the rank of a science, until men of science have perfected both the measures of heights by barometrical observations, and the determination of mean temperatures; or what is more important for the development of vegetation, the determination of the differences between the temperature of summer and winter and between that of day and night" (Humboldt 1816a, p. 446, original emphasis).

Humboldt's geography of plants was about measuring and did not contain philosophical musings or a historical synthesis within a systematic method. Here we return to the "age of monopolies": Humboldt did have a monopoly on scientific measurements and data. A botanist with a modest budget was unable to venture far to acquire the latitudes, temperatures and altitudes of flora and vegetation in remote places, like South America. Large sums of data would only be available to those like Humboldt who had used a third of his inheritance to travel (de Terra 1955, p. 198, footnote 1), collect data and publish, therefore claim leadership of a field that respects "the altitudes which spontaneous plants reach in different climates". But collecting data without a rigorous systematic method leaves one to wonder whether the geography of plants had two very different aims:

> Candolle carefully distinguished the larger concerns of phytogeography from what has come to be known as ecology (Browne 2002, p. 1).[21]

and,

> ... Alexander von Humboldt's plant geography was indeed centrally concerned with vegetation – its character, distribution and relation to environmental parameters – and not solely or primarily with individual plants or species (Nicolson 1996, p. 289).

These two different fields are a result of different influences within botanical circles.

## Humboldt and de Candolle's Methods Compared

Humboldt had a privileged upbringing. After graduating from the University of Frankfurt an der Order, Alexander von Humboldt formed a friendship with botanist Carl Ludwig Willdendow his mentor and the first German to discuss plant geography (Nicolson 1987, p. 174). Next Humboldt attended the prestigious University of Göttingen in 1789, where he was mentored by Johann Friedrich Blumenbach (1752–1840). During his time at Göttingen, he befriended Georg Forster who introduced him plant geography (Hartshorne 1958, p. 99). Later Humboldt presented Forster "... the first outline of a Geography of Plants in 1790" (Humboldt in Lomolino

---

[21]Browne may have confused Augustin Pyramus de Candolle with his son Alphonse, who published *Géographie Botanique Raisonnée* (de Candolle 1855), a work that is often and possibly erroneously attributed as the "birthplace" of ecology, to the detriment of foundational ecological texts like Schouw (1823) and Meyen (1846) as well as Johannes Eugenius Bülow Warming's *Ecology of plants* (Warming 1895) and Andreas Franz Wilhelm Schimper's *Plant geography on a physiological basis* (Schimper 1898). As Raup clearly points out "Ecological plant geographers go back to Humboldt, Schouw and Grisebach for the classic foundation of their view. This seems to be due to the fact that these students broke away from the purely floristic idea of plant geography and set up descriptive units based upon pure form and community structure [...] We have already seen, however, that in spite of these changes, the processes of logic had, for the older men, remained identical with those of the floristic geographers" (Raup 1942, pp. 330–331). For a discussion on the Humboldtian connection see Nicolson (1996).

et al. 2004, p. 49). Ironically, Stromeyer and Humboldt graduated from the same university, albeit almost 10 years apart, they were mentored by different people.

The geographical relationship between Candolle and Humboldt was not as close. Candolle drew inspiration for his botanical geography from the works of the natural classifications of taxonomic groups, such as Antoine Laurent de Jussieu (1748–1836), and not vegetation. In contrast, Humboldt's geography of plants was about finding the physical or geographical characteristics that influenced the distribution of vegetation types. Candolle's approach, first outlined in the 3rd edition of the *Flore française* (Lamarck and Candolle, 1805),[22] looked at the distribution of floras, actual assemblages of species, based on certain physical or geographical factors. The difference between these approaches was how each dealt with geography and classification. Candolle saw geography as distribution over time, while he took at highly systematic approach to classification. In contrast, Humboldt did not look at floras, but at vegetation within an ahistorical context. His approach was not systematic. Rather it relied on measurements of physical geography, such as temperatures, latitudes and so on. The anonymous letter (like the lectures on geography by Kant) reveals two plant geographies: physical or vegetation plant geography and, the historical or taxonomic geography of plants.

Given Humboldt and Candolle's varying definitions for botanical geography, only Humboldt questions the meaning of geography, its origin or relevance to his physiognomy that is an arbitrary or aesthetic physical description. Candolle however has a far more pragmatic approach. He like many taxonomists wanted to 'solve' botanical distributions through classification and historical process, linking, as it would seem, his work back to that of Linnaeus. In any case, neither Humboldt nor Candolle seem to resolve the issue highlighted by Raup, namely to reach 'a resolution of the complex interplay of influences in a geographic area'. In order to do so, it is important to go back to Kant's *Physical Geography* to trace the influences that lead to Humboldt's physiognomy and Candolle's belated grudge.

But what of Humboldt's legacy? Indeed he was the first to practise plant geography (Stromeyer's phytogeography), while others like Stromeyer simply wrote anthologies (historical plant geographies), classifying and comparing the discoveries and the work of other within a handbook. It is hard to imagine why Humboldt saw this as an affront to his own geography of plants. Anthologies are not only useful, but they provide the basis for teaching plant geography and introducing plant distributions to a new audience. But Humboldt was a self-funded and privately educated naturalist, not a specialised academician with a doctorate in plant geography. Humboldt saw the geography of plants as a far more practical field, with data, methodology and analysis. Seen in this way, one may start to appreciate Humboldt's point-of-view: why should an anthology, derived from the descriptive taxonomic treatments like those of Willdenow and Linnaeus, be allowed to surpass

---

[22]De Candolle was the sole contributor to the third edition of *Flore française* (Lamarck and de Candolle 1805): "The young botanist was entrusted with the task of bringing out a new edition of Lamarck's *Flore française*" (Drouin 2001, p. 256; see also Stevens 1994, p. 205).

an empirical work based on vast amounts of data which was years in the making? After all, Stromeyer's *Specimen* appeared while Humboldt and Bonpland were out in the field busily collecting, measuring and practicing plant geography. Even before he had encountered Stromeyer's *Specimen*, Humboldt's determination to break from traditional taxonomic treatments is obvious in his *Essai*,

> Botanists usually direct their research towards objects that encompass only a very small part of their science. They are concerned almost exclusively with the discovery of new species of plants, the study of their external structure, their distinguishing characteristics, and the analogies that group them together into classes and families (Humboldt and Bonpland 2009, p. 64).

Implementing a "geography of plants" did pay off. Humboldt had clearly retained his place as the founder of plant geography while Stromeyer was clearly the loser. For instance, Wulff (1943) chose Schouw's usage of 'historical plant geography', to that of Stromeyer, while other historical treatments, notably by Hofsten (1916) and Schmithüsen (1985), ignore Stromeyer's contribution altogether. For instance, Wittwer (1860) sums up Stromeyer's contribution in a sentence,

> ... Stromeyer provided a plan of how plant geography should be arranged, a catalogue of [botanical] works [...] is concerned more with the concept than with plant geography itself (Wittwer 1860, p. 215).[23]

But the number of such catalogues or anthologies bloomed, using a style and layout similar to that of Stromeyer (e.g., Cain 1944, Grisebach 1872, Meyen 1846, Schouw 1823, Wulff 1943), namely a preface and dedication, an introduction to the work, followed by the former distribution of taxa or vegetation types, then by descriptions of modern distributions, usually based on empirical work, travels or discoveries made by other people. These anthologies differ in what explanations or processes they adopt to explain such distributions. These are simply worked into each section. Most anthologies begin with an introduction, followed by a goal or aim and definition (usually of the field of study). The majority of the text describes the distribution of taxa or vegetation types and their relationship to geographical processes, such as substrate and weather. Given this, it could be possible to argue that Stromeyer had in fact created a tradition of classification and anthology within plant geography, similar to that of Zimmermann's anthology of animal distribution (Zimmermann 1777), a practise that still exists in modern plant geography texts (e.g., Moreira-Muñoz 2011, Stott 1981).

---

[23]Wittwer's claim is ironic as Stromeyer set up the first teaching laboratory for chemistry. In fact, "... one laboratory was far ahead of the others in terms of scale, quality of teaching, and reputation: the laboratory of Friedrich Stromeyer at Göttingen. Not only did many pharmacists receive their practical training here, but also countless future state physicians, agronomists, mineralogists and mining officials. Students came from all over Germany to study analytical chemistry at Göttingen" (Homburg 1999, p. 15; see also Lockeman and Oesper 1953, p. 202).

## The Geographies of Kant and Humboldt

At the time when Kant presented his lectures on 'Physical Geography' at the University of Königsberg (1756–1796),[24] geography was already an emerging field in the eighteenth century that had several protagonists (see Withers 2006).[25] Kant's lectures discussed knowledge and how it related to the observable world. Like many naturalists of the time, there was a burning desire to legitimise geography as a unified and respectable scientific field, by arguing for its foundations and classifying its varying aspects. For Kant, physical geography was to all intents and purposes a descriptive field that acts as a basis for history. That is, geography acts as a starting point for any history:

> The name geography therefore designates a description of nature, and at that of the whole earth. Geography and history fill up the total span of our knowledge; geography namely that of space, but history that of time.
> 
> We ordinarily assume that there is an old and a new geography, because geography has existed at all times. But which came first, history or geography? The latter is the foundation of the former, because occurrences have to refer to something. History is a never relenting process, but things change as well and result at times in a totally different geography. Geography therefore is a substratum. Since we have an ancient history, so naturally we must have an ancient geography (Kant, edited by Rink in May 1970, pp. 261–262; see also the translation of Reinhardt 2012, pp. 450–451[26]).

Kant's treatment of history is interesting as it places physical geography "as a general compendium of nature, [that] is not only the foundation of history but of all other geographies", by which he means mathematical, moral, political, commercial and theological geography. This division between history and geography[27] also appears between Linnaeus' taxonomy and a classification of things in the world:

> The classification of knowledge by concepts is the logical, that by time and space the physical classification. Through the former we obtain a system of nature (*Systema Naturae*), for example that of Linnaeus; through the latter a geographical description of nature [...] The system of nature is, so to speak, a register of the whole, in which I place each thing in

---

[24] For a detailed historical account of Kant's lectures in geography see Elden (2009) and May (1970). For a modern translation of Kant's lectures in geography see Reinhardt (2012).

[25] These geographies included both 'physical' and 'cultural' aspects, such as Mentelle's *Eléments de géographie* (Mentelle 1758) and Meletios' Γεωγραφία παλαιά καὶ νέα [*Geography, old and new*] (Meletios 1728) respectively.

[26] Olaf Reinhardt's translation differs from May: "We usually refer to the geography of the past and of the present, for geography has existed at all times. But which came first, history or geography? The latter is a prerequisite for the former, because events necessarily take place with reference to something. History is a continuous progression, but things, too, change, and give an entirely different geography at particular times. Geography is thus the foundation [of history]. If we have ancient history, naturally we must also have ancient geography" (Reinhardt 2012, pp. 450–451).

[27] According to Kant the difference between what we observe and what is told to us via a narrative "History is a narrative, but geography is a description. Therefore we may have a description of nature, but not a history of nature" (Kant in May 1970, p. 260).

its proper class, even though they are to be found in far-flung regions of the earth (Kant in May 1970, p. 259; see also the translation of Reinhardt 2012, pp. 448[28]).

In other words, classification, in the case of Linnaeus' *Systema Naturae*, is incomplete and logical (abstract or artificial), where as an actual classification based on the physical descriptions of say plants and their current environs would be a more natural or physical classification that is based on observation. If Rink's transcriptions rings true, then Kant had by the mid eighteenth century already identified the rift that is to plague nineteenth century geography, namely the battle between logical versus physical classification.

Humboldt never heard of Kant's lectures (prior to Rink's 1802 publication).[29] This may not discount Humboldt from ever encountering Kant's ideas on physical geography, but it is more likely that Abraham Gottlob Werner, Humboldt's teacher contributed a large part of Humboldt's thinking (see Hartshorne 1958, pp. 100–101).[30] Nonetheless, as exciting as it was to the young 24 year old Humboldt, his life's work was not to leave the geography of plants 'untouched'.

Kant considered geography to be a description of a nature devoid of any historical context. That is, geography fulfils a system of nature, which attempts to order objects. "In the existing so-called system of this type, the objects are merely put beside each other and ordered in sequence on after the other [...] We can call both history and geography, at the same time, a description, but with the difference that the former is a description of time while the other is a description of space" (Rink, in May 1970, p. 260).

Rink's notes go further to show Kant's concept of geography as a separate field to history "Because the history of nature is no younger than the world itself, we cannot vouch for the accuracy of our reports, not since the invention of writing. And what an immeasurable and probably far greater time lies beyond what is presented to us in recorded history" (Rink, in May 1970, p. 261). In other words, we can survey land and pin-point places, because it we are able to go there and measure them. In history, however, we rely on written accounts, on unobservable events that we cannot vouch for, measure or witness. In Rink's notes, Kant uses the natural history of dogs as an example, in which he discusses how "various breeds of dogs have come from a same root, and what changes have occurred in them in various countries and climates" (Rink, in May 1970, p. 261). We can speculate what we know about the natural history of dogs, but we know for example, where particular dogs are found. This highlights the difference between what can be observed and

---

[28]Reinhardt's translation differs again from that of May, "Division of knowledge according to concepts is logical; according to time and space it is physical. By means of the former, we obtain a system of nature (*Systema naturae*), as for example that of Linnaeus. With the latter, we obtain a geographical description of nature [...] The *Systema naturae* is, as it were, a kind of register of the whole, wherein I situate all things, each in the class to which it belongs, even if on earth they are to be found in widely separated areas" (Reinhardt 2012, p. 448).

[29]But as Hartshorne (1958, p. 101) points out, Humboldt may have read Rink's 1802 edition of Kant's lectures, leading him to make similar remarks as Kant in his *Kosmos* (Humboldt 1849).

[30]Geognosy, a term Humboldt uses in his *Floræ Fribergensis*, describes a new field of enquiry, namely historical geology.

how it relates to other observable things spatially versus what can be extrapolated from written sources. Also there is the issue of context. In a newspaper, we may read of a war in a far-flung place. To the reader not knowledgeable in geography, the news story has no significance. "And yet, for instance, when the news tells something about the course of a ship through the polar sea, this is a most interesting matter because the now almost impossible hope of discovering a passage through the polar sea could bring to Europe the most important changes" (Rink, in May 1970, p. 262). For Kant history cannot work without a systematic geography (e.g., Paris is located in Europe) or classification of places and our relation to them. Like Kant, Humboldt's geography looks for associations between objects and process, such as the distribution of plants and climate, however, where it differs is in its lack of a system or physical classification.

The importance of a system or a comparative method methodology is vital in taxonomy. Most naturalists were taxonomists or used a type of taxonomy, for instance, the crystal or chemical groups. Classification underpins all comparative science, so it is not surprising that Kant identifies two classifications:

> Concerning a plan for order, we have to designate a particular place to all our knowledge. We can classify our knowledge of experience either according to concepts or according to the time and place where it is actually found.
>
> The classification of knowledge by concepts is the logical, that by time and space the physical classification. Through the former we obtain a system of nature (*Systema Naturae*), for example that of Linnaeus; through the latter a geographical description of nature [...] The system of nature is, so to speak, a register of the whole, in which I place each thing in its proper class, even though they are to be found in far-flung regions of the earth. On the other hand, following the physical classification, things are observed according to the places, which they occupy on the earth. The system gives position in the classification. But the geographical description of nature shows us the place in which every object on the earth is really to be found [...] In general, we consider here the scene of nature, the earth itself and the regions where things are really to be found. In the system of nature, however, the question is not about native places but about similarities of form (Kant in May 1970, p. 259, see also the translation of Reinhardt 2012, p. 448).

Humboldt's plant geography does not fit into Kant's classification.[31] Plant geography as Humboldt conceived it, was not comparative. Rather he created a general cosmology of the interrelationships between plants and the universe.

---

[31] In his *Kosmos*, Humboldt echoes Kant, stating that, "[w]e would first distinguish between the physical *history* and the physical *description* of the world". Humboldt however makes the mistake of confusing history with geography "But if we would correctly comprehend nature, we must not entirely or absolutely separate the consideration of the present state of things [Kant's geography] from that of the successive phases through which they have passed [Kant's history]. We cannot form a just conception of their nature without looking back on the mode of their formation. It is not organic matter alone that is continually undergoing change and being dissolved to form new combinations. The globe itself reveals at every phase of its existence the mystery of its former conditions (Humboldt 1864, p. 54). This geological perspective, possibly derived from his mentor Werner, would fall into Kant's system of nature. In 1807 however, Humboldt's attitude is purely geographical.

Humboldt's plant geography was unique as it differed from that of Linnaeus, Willdenow and Candolle. Perhaps this is why he dismissed each as being part of "a science that up to now exists in name only, and yet is an essential part of general physics" (Humboldt and Bonpland 2009, p. 64).

Note the term 'general physics'. Herein lies Humboldt's cosmology, a field that describes the interrelationships of physical phenomena: "I have attempted to gather in one single tableau the sum of the physical phenomena present in equinoctial regions, from the sea level of the South Sea to the very highest peak of the Andes". This tableau contains:

> The vegetation; The animals; Geological phenomena; Cultivation; The air temperature; The limit of perpetual snow; The chemical composition of the atmosphere; Its electrical tension; It barometric pressure; The decrease in gravity; The intensity of the azure color of the sky; The weakening light as it passes through the strata of the atmosphere; The horizontal refractions, and the temperature of boiling water at various altitudes (Humboldt and Bonpland 2009, p. 78).

Candolle's 'plant geography' however, aimed to compare areas based on a limited number of comparable characteristics,

> From the preceding considerations, I believe that in a given country, such as France, the causes that determine the plant region [habitation] could be reduced to three:
>
> 1. Temperature, as determined by distance from the equator, height above sea level and southern or northerly exposure.
> 2. The mode of watering, which is more or less the quantity of water that reaches the plant. The manner by which water is filtered through the soil and the matter that is dissolved in the water which may or may not be harmful to the growth of the plant.
> 3. The degree of soil tenacity or mobility (Candolle in Ebach and Goujet 2006, p. 766).

Candolle's method allows two areas to be compared based on three properties. Humboldt's method is a general description of the interactions between plants and their surrounding physical phenomena. Each area will have their own unique physical phenomena not necessarily found in other regions (e.g., compare a desert region with a tropical rainforest).

While we can justify Humboldt and Candolle's methods as geographical based because they concern themselves about "the earth itself and the regions where things [i.e., plants] are really to be found", namely a physical classification, only the latter is comparative. This distinction is by no means trivial. Humboldt and Candolle were attempting two very different things leading to two very different scientific pursuits. Humboldt's method stems from Werner's Geonosy (*Erdkunde*) while Candolle's method, like Willdenow's, is based on Linnaeus' system of nature.

In hindsight, Humboldt's method seems to be the most comprehensive and complete. Humboldtian plant geography for a twenty-first century scientists must seem like a Herculean subject. Geognosy, or more precisely Earth science, is exactly what Humboldt was describing, *sans* animals. The same could be said for human geography, a scientific field that investigates every conceivable impact on human geography, therefore plant geography, as Humboldt conceived it, views it from the perspective of plants. Regardless of these vast differences between the plant geography of Humboldt, Candolle and Willdenow, historians still confuse both. For instance, Nordenskiöld (1936) mentions that Linnaeus started systematic plant geography and Humboldt founded a morphological plant geography, which he equates with ecology.[32] Of course it is impossible to equate Humboldt's work with later nineteenth and twentieth century ecology, which asks very different questions and is limited to certain environmental and evolutionary processes within populations. Humboldt's original plant geography and his later *Kosmos* is a forgotten field, more similar to human geography than to modern day ecology or biogeography. Within his *Kosmos* Humboldt glosses over plant geography, stating his claim to a field "which no has as yet been given" and one that "Menzel,[33] in an inedited work on the flora of Japan, accidentally made use of the term *geography of plants*; and the same expression occurs in the fanciful but graceful work of Bernardin de St. Pierre, *Etudes de la Nature*" (Humboldt 1858, p. 347). Humboldt continues:

> ... [a] scientific treatment of the subject began, however, only when the geography of plants was intimately associated with the study of the distribution of heat over the surface of the earth, and when the arrangement of vegetable forms in natural families admitted of a numerical estimate being made of the different forms which increase or decrease as we recede from the equator towards the poles, and of the relations in which, in different parts of the earth, each family stood with reference to the whole mass of phanerogamic indigenous plants of the same region (Humboldt 1858, pp. 347–348).

Humboldt was of course referring to himself, giving no mention to Candolle, Willdenow or Stromeyer (see below). Rather Humboldt refers to Schouw, Endlicher and Unger ("*The history of plants*, which Endlicher and Unger have described in a most masterly manner, I myself separated from the *geography of plants* half a century ago" [Humboldt 1858, footnote on p. 340]), a new generation of naturalists influenced by his work. This may at first seem rather petty of such an iconoclastic historical figure to take credit for a field that he shares with others. But Humboldt had a vision for a wholly descriptive field that catalogues the interaction between surface phenomena and organic life "And it is by these links that *the geography*

---

[32]Nordenskiöld's misconceives Humboldt's plant geography is several ways: "The greatest service rendered to biology by Humboldt, however, was his creation of vegetable geography [...] The whole of this conception of plant life and this grouping of its individual components according to common conditions of life, instead of according to the nomenclature of species, represent a new idea [...] he created a new field of research, which was cultivated at a later period with great success (Nordenskiöld 1936, pp. 315–316)." Humboldt's work is too far to ranging to be lumped into any later nineteenth century field, like ecology or biogeography. Rather Humboldt's disciples such as Schouw, Grisebach and Schimper, who had selective picked ideas from the *Essai*, could be said to have formulated nineteenth century ecology (see Nicholson 1996).

[33]Christian Mentzel was re-discovered by Lesser (1751, p. 321) "some years ago" when "a rag-and-bone lady [Trödelfrau] brought me a battered book in folio to sell and, it was by the famous D. Christani Mencelii, *index nominum plantarum universalis* [1682]". Lesser continues "it would be better if stated that the fatherland of plants in length and breadth, rather than: these plants occur in America, in India [...] Herein it is useful for every *area geographico-botanica* to have a *floram* ..." (Lesser 1751, p. 323, footnote *).

*of organic beings – of plants and animals –* is connected with the delineation of the inorganic phenomena of our terrestrial globe" (Humboldt 1858, p. 341). Neither Candolle nor Willdenow had such grand ambitions as Humboldt. But why should they? Plant taxonomists interested in classifying species and the areas in which they occur. While Humboldt tackled a general description and history or the Earth, other plant geographers were looking at comparing areas for classification. The units of comparison were the species and the factors that affected their distribution.

Humboldt's *Essai,* descriptive in nature and devoid of taxonomy, contrasts starkly to Willdenow's *Grundriss der Kräuterkunde,* taxonomic in nature and historical. Willdenow was the first in a long line of German botanists to write large introductory texts to plant geography. What makes *Grundriss* so interesting is that it attempts to update systematic botany. Rather than group plants by their behavioural and physical interactions with the environment, Willdenow classified them in the Linnaean sense by physiognomy or form. These two classifications, namely *vegetative* and *floristic* respectively, are found throughout the plant geography literature. Unbeknownst to late eighteenth century naturalists like Willdenow[34] and Humboldt, vegetative classifications are purely artificial, depending on what association the classifier makes at the time (e.g., compare a 'rainforest' in Australia and Brazil). Floristic classifications are dependent on taxonomy (e.g., compare the Central American Avocado with a the southeast Asian Cinnamon tree). Jungles share a description ("an overgrown area of land"), whereas natural taxonomic groups share a common history (avocados and cinnamon are both members of the Family Lauraceae). Comparing across taxonomies gives greater emphasis to history than comparing across environmental conditions. The latter sets a precedent, it is comparable no matter which group you study and that certain questions could be asked:

> From where comes this vast diversity of plants that our Earth produces? (Willdenow 1810, p. 485, my translation).

The first step in answering such a question is to find out where individual plants are distributed:

> The northern flora is distributed across Denmark, Sweden and Russia, as well as a part of England (Willdenow 1810, p. 514, my translation).

Once we have determined our distributions, we can ask further questions,

> When Mr Forster found *Pinguicula alpina, Galium Aparine, Armeria vulgaris,* and *Ranunculus lapponicus* in the Tierra Del Fuego, it must have seemed difficult to explain how these plants reached the furthest corners of the globe [ ... ] I have no doubt, that among the numerous plants, which arise on our globe, there are not some that are largely flexible enough to withstand all climatic conditions, like such animals as Man, the dog and the pig, which are found in all zones (Willdenow 1810, p. 508, 519).

---

[34]Willdenow " ... neither loved Paris, nor the more modern French plant systematics" meaning that 'natural groups' in the sense of Geoffroy Saint-Hilaire's *Theory of Analogs* (homology) may have been anathema to the Prussian botanist.

This leads to explanations that attempt to explain these distributions:

> Plants like animals are bound to certain distributions. Different types from warm climes, could adapt to our colder climate (Willdenow 1810, p. 510).

Willdenow's *Grundriss* "was the first to lay out a systematic set of principles of the geography of plants. These severed as a springboard for Humboldt's synthesis" (Jackson, in Humboldt and Bonpland 1807, p. 251). Unlike Willdenow however, Humboldt attempts to explain a natural history of how plants are distributed, without context to any previous work "botanists direct their research towards objects that encompass only a very small part of their science" (Humboldt and Bonpland 1807, p. 65). A dismissal like this, one would think, could only come from someone who is not directly involved in a field. Conversely, Willdenow (1810) presents an anthology of botanists since in his "History of Science", an invertible 'who's who' in botany up to 1810, 74 pages in total covering eight 'epochs' starting with the ancient Greek Aesoulap.[35] Humboldt is listed but not his work on the geography of plants. Rather Willdenow lists Humboldt as the Prussian Chamberlain [Kammerherr][36] who "with a hopeful Aimé Bonpland travelled through a large part of the Spanish territories in America, and brought back with them a trove of natural treasures" (Willdenow 1810, p. 548). Strangely there is no mention of Humboldt's geography of plants in Willdenow's *Grundriss* at all.[37] Here we return to Raup and his claim that the nineteenth century, "has never hoped for more than an approximate solution to the problem of ultimate causation".

Given the diverse nature of the geographies of de Candolle and Humboldt, their different aims and goals, one would agree with Raup's "approximate solution". Hartshorne's unification is one that is still echoed by biogeographers today, even though there is no common origin. De Candolle was trained as a taxonomist in a systematic methodology, in which the goals were to find a natural classification, one that would make sense of taxonomy. Humboldt had no such training, even though he was tutored by the likes of Willdenow, had also sought to find a natural order, but one of causation. Where both de Candolle and Humboldt overlap however is in their notion that Earth processes, like temperature, soils and hydrology, affect the geography of plants. Their differences are how they classified plants, either taxa or vegetation and, how they defined geography. While de Candolle was concerned

---

[35] Humboldt seems to be missing from Willdenow's 'History of the Science' section in later German and English editions up to 1831.

[36] In the English edition (Willdenow 1811) Humboldt is listed as the "Chief Counselor of mines in Prussia" who had "much contributed to the knowledge of subterraneous plants". There too is no mention of his geography of plants.

[37] Willdenow's text also includes a section on the History of Plants defined as "the influence of climate on vegetation; the changes which it is probable plants undergo from the revolutions of our globe; their dispersion over its surface; their migrations; and, lastly the means pursued by Nature for their preservation" (Willdenow 1811, p. 407). Here too Humboldt's geography of plants is not mentioned.

about taxic distributions, Humboldt looked at the broader physical geography with less attention to actual taxonomy. In later years, this conceptual divide created two very different plant geographies heralding two very different origins.

## References

Anonymous. (1820). Einige Bemerkungen über zwei, die Pflanzengeographie betreffende Werke des Hrn. v. Humboldt. Im einem Schreiben an den Hofrath Scharder. *Jahrbücher der Gewächskunde, 1*, 6–56.
Anonymous. (1821). III. Observations respecting the Geography of plants Addressed to the Editor of the *Quarterly Journal of Science*, & c. *Quarterly Journal of Science, Literature and Art, 19*, 236–267.
Anonymous. (1825). II. Essai Géognostique sur le Gisement des Róches dans les deux hémisphéres, par Alexandre de Humboldt. *Quarterly Journal of Science, Literature and Art, 10*, 306–327.
Asimov, I. (1972). *Asimov's biogeographical encyclopedia of science and technology*. London: Pan books.
Bauer, J. (1852). *Lives of the brothers Humboldt, Alexander and William*. London: Ingram, Cooke and Co.
Beer, G. (1999). Eine Idee von der Geographie der Pflanzen – oder 'Im Schatten Alexander von Humboldts'. Dr. med. Friedrich Stromeyer und seine Briefe aus Frankreich 1801–1802 an seine Familie in Göttingen. Ein Göttinger erzählt von Paris und von seiner botanisch-mineralogischen Reise in den Pyrenäen. *Museum der Göttinger Chemie Museumsbrief, 18*, 1–48.
Beer, G. (2006). Friedrich Stromeyer und sein Göttinger Schüler Edward Turner, 1828 der erste Chemiker am University College, London. *Museum der Göttinger Chemie Museumsbrief, 25*, 3–13.
Bowen, M. (1981). *Empiricism and geographical thought. From Francis Bacon to Alexander von Humboldt*. Cambridge: Cambridge University Press.
Browne, J. (2002). *Augustin-Pyramus de Candolle. Encyclopedia of life sciences*. New York: Wiley.
Buffon, G. L. L. C. D. (1749–1789). *Histoire naturelle: générale et particulière, servant de suite à l'histoire des animaux quadrupèdes* (36 vols.). Paris: L'Imprimerie Royale.
Cain, S. A. (1944). *Foundations of plant geography*. New York: Harper and Brothers.
Castrillon, A. (1992). Alexandre de Humboldt et la géographie des plantes. *Revue d'histoire des sciences, 45*, 419–433.
de Candolle, A. P. (1805). Explication de la carte Botanique de la France. In J. B. P. A. de Lamarck & A. P. de Candolle (Eds.), *Flore française, ou descriptions succinctes de toutes les plantes qui croissent naturellement en France, disposées selon une nouvelle méthode d'analyse, et précédées par un exposé des principes élémentaires de la botanique* (3rd ed.). Paris: Desray.
de Candolle, A. P. (1813). *Théorie Élémentaire de la Botanique, ou Exposition des Principes de la Classification Naturelle et de l'Art de Décrire et d'Etudier les Végétaux*. Paris: Déterville.
de Candolle, A. P. (1817). Mémoire sur la géographie des plantes de France, considérée dans ses rapports avec la hauteur absolue. *Mémoires de Physique et de Chimie de la Société d'Arcueil, 3*, 262–298.
de Candolle, A. P. (1820). Essai élémentaire de géographie botanique. *Dictionnaire des Sciences Naturelles* (Vol. 18, pp. 1–64). Paris: F. Levrault.
de Candolle, A. L. P. P. (1855). *Géographie botanique raisonnée*. Paris: Masson.
de Candolle, A. L. P. P. (1862). *Mémoires et souvenirs*. Geneva: Cherbuliez.
de la Roquette, J.-B. M. A. D. (1865). *Humboldt: correspondance, scientifique et littéraire recueillie*. Paris: E. Ducrocq.

# References

de Terra, H. (1955). *Humboldt: The life and times of Alexander von Humboldt, 1769–1859*. New York: Knopf.
Detwyler, T. R. (1969). Humboldt's essay on plant geography: Its relevance today. *Michigan Academician, 3–4*, 113–122.
Drouin, J.-M. (1998). Augustin-Pyramus de Candolle. In P. Acot (Ed.), *The European origins of scientific ecology, Volume 1* (pp. 359–422). Amsterdam: Overseas Publishers Association.
Drouin, J.-M. (2001). Principles and uses of taxonomy in the works of Augustin-Pyramus de Candolle. *Studies in History and Philosophy of Science Part C, 32*, 255–275.
Drouin, J.-M. (2004). Introduction. In J.-D. Candaux & J.-M. Drouin (Eds.), *Augustin-Pyramus de Candolle – Mémories et souvenirs (1778–1841)* (pp. 1–35). Geneva: Georg.
Ebach, M. C., & Goujet, D. (2006). The first biogeographical map. *Journal of Biogeography, 33*, 761–769.
Elden, S. (2009). Reassessing Kantłs geography. *Journal of Historical Geography, 35*, 3–25.
Engelmann, G. (1966). Carl Ritter 'Sech Karten von Europa' Mit einer Abbildung. *Erdkunde, 10*, 104–110.
Giraud-Soulavie J.-L. (1783). *Histoire naturelle de la France méridionale, Volume 1. Chez J.F. Quillau, Mérigot l'aîné*, Belin: Paris.
Gmelin, J.F. (1801). Stück [Article] 170. *Göttingischen Anzeigen von gelehrten Sachen,1801*, 1696.
Goethe, J. W. (1883). *Conversations of Goethe with Eckermann and Soret*. London: George Bell & Sons.
Gray, A. (1889). *Scientific papers of Asa Gray*. Boston: Houghton.
Grisebach, A. H. R. (1872). *Die Vegetation der Erde nach ihrer klimatischen Anordnung*. Leipzig: Wilhelm Engelmann.
Hartshorne, R. (1939). The nature of geography: A critical survey of current thought in the light of the past. *Annals of the Association of American Geographers, 29*, 173–412.
Hartshorne, R. (1958). The concept of geography as a science of space, from Kant and Humboldt to Hettner. *Annals of the Association of American Geographers, 48*, 97–108.
Homburg, E. (1999). The rise of analytical chemistry and its consequences for the development of the German chemical profession (1780–1860). *Ambix, 46*, 1–32.
Jackson, S. T. (2009). Introduction: Humboldt, ecology, and the Cosmos. In A. von Humboldt & A. Bonpland (Eds.), *Essay on the geography of plants* (edited by S. T. Jackson) (pp. 1–52). Chicago: University of Chicago Press.
Leotard, S. (1893). *Lettres Inedites de Moquin-Tandon a Auguste de Saint-Hilaire*. Clermont-L'Herault: Saturnin Leotard.
Lesser, F. C. (1751). Nachricht von der Dr. Menzel angegebenen botanischen Geographie. *Physikalische Belustigungen, 1*, 321–327.
Link, H. F. (1820). Vorrede. *Jahrbücher der Gewächskunde, 1*, i–iv.
Lockemann, G., & Oesper, R. E. (1953). Friedrich Stromeyer and the history of chemical laboratory instruction. *Journal of Chemical Education, 30*, 202–204.
Lomolino, M. V., Sax, D., & Brown, J. H. (2004). *Foundations of biogeography: Classic papers with commentaries*. Chicago: Chicago University Press.
May, J. A. (1970). *Kant's concept of geography and its relation to recent geographical thought*. Toronto: University of Toronto Press.
Meletios, M. (1728). *Geography old and new*. Venice: Glykis.
Mentelle, E. (1758). *Eléments de géographie, contenant les principales divisions des quatre parties du monde, avec de courtes explications fur chacune d'elles, une description abrégée de la France, à l'usage des commençants*. Paris: Barrios.
Mentzel, C. (1682). *Index nominum plantarum universalis multilinguis*. Berlin: Ex Officina Rungiana.
Meyen, F. J. F. (1846). *Outlines of the geography of plants*. London: Ray Society.
Moreira-Muñoz, A. (2011). *Plant geography of Chile*. Dordrecht: Springer.
Nelson, G. (1978). From Candolle to Croizat: Comments on the history of biogeography. *Journal of the History of Biology, 11*, 269–305.

Nicolson, M. (1987). Alexander von Humboldt, Humboldtian science and the origins of the study of vegetation. *History of Science, 25*, 167–194.
Nicolson, M. (1996). Humboldtian plant geography after Humboldt: The link to ecology. *The British Journal for the History of Science, 29*, 289–310.
Nordenskiöld, E. (1936). *The history of biology: A survey.* Translated from the Swedish by Leonard Bucknall Eyre. New York: Tudor.
Raup, H. (1942). Trends in the development of geographic botany. *Annals of the Association of American Geographers, 32*, 319–354.
Reinhardt, O. (2012). Physical geography (1802). In E. Watkins (Ed.), *Immanuel Kant: Natural Science* (pp. 426–679). Cambridge: Cambridge University Press.
Schimper, A. F. W. (1898). *Pflanzen-Geographie auf Physiologischer Grundlage.* Jena: Gustav Fischer.
Schmithüsen, J. (1985). Vor- un Frühgeschichte der Biogeographie. *Biogeographica, 20*, 1–166.
Schouw, J. F. (1823). *Grundzüge einer allgemeinen Pflanzengeographie.* Berlin: Reimer.
Schrader, H. A. (1800) [1801]. Commentatio inauguralis sistens historiae vegetabilium geographicae specimen. *Journal für die Botanik, 1*, 484–393.
Stearn, W. T. (1960). Humboldt's Essai sur la géographie des plantes. *Journal of the Society for the Bibliography of Natural History, 3*, 351–357.
Stevens, P. F. (1994). *The development of biological systematics: Antoine-Laurent de Jussieu, nature, and the natural system.* New York: Columbia University Press.
Stoddard, R. H. (1859). *The life, travels and books of Alexander von Humboldt.* New York: Rudd and Carleton.
Stott, P. (1981). *Historical plant geography.* London: George Allen and Unwin.
Stromeyer, F. (1800). *Commentatio inauguralis sistens historiae vegetabilium geographicae specimen.* Göttingen: Heinrich Dieterich.
Thomson, T. (1830). *The history of chemistry.* London: Colburn and Bentley.
von Hofsten, N. G. E. (1916). Zur älteren Geschichte des Diskontinuitätsproblems in der Biogeographie. *Zoologische Annalen Zeitschrift für Geschichte der Zoologie, 7*, 197–353.
von Humboldt, A. (1790). *Mineralogische Beobachtungen über einige Basalte am Rhein.* Braunschweig: Schulbuchhandlung.
von Humboldt, A. (1793). *Florae Fribergensis specimen.* Berlin: H. A. Rottmann.
von Humboldt, A. (1815). Prolegomena. In A. von Humboldt & A. Bonpland (Eds.), *Nova genera et species plantarum, quas in peregrinatione ad plagam aequinoctialem orbis novi collegerunt, descripserunt, partim adumbraverunt.* Paris: Apud Gide Filium.
von Humboldt, A. (1816a). XCIII. On the laws observed in the distribution of vegetable forms. *Philosophical Magazine Series 1, 47*, 446–453.
von Humboldt, A. (1816b). Sur les lois que l'on observe dans la distribution des formes végétales. *Annales de chimie et de physique, 1*, 225–239.
von Humboldt, A. (1823). *Essai Géognostique sur le Gisement des Róches dans les deux Hémisphères.* Strasbourg: Chez F. G. Levrault.
von Humboldt, A. (1849). *Cosmos: Sketch of a physical description of the universe* (Vol. 2). London: Longman, Brown, Green and Longmans.
von Humboldt, A. (1858). *A sketch of the physical description of the universe* (Vol. 1). New York: Harper & Brothers.
von Humboldt, A. (1864). *Cosmos: Sketch of a physical description of the universe* (Vol. 1). London: Henry G. Bohn.
von Humboldt, A., & Bonpland, A. (1807). *Voyage de Humboldt et Bonpland. Première partie. Physique Générale, et relation historique du voyage* (Premier Volume, Contenant Essai sur la Géographie des plantes, accompagné d'un Tableau physique des régions équinoxiales, et servant d'introduction à l'Ouvrage). Paris: Chez Fr. Schœll.
von Humboldt, A., & Bonpland, A. (2009). *Essay on the geography of plants.* Chicago: University of Chicago Press.
von Wilcke, G. (1967). Der Chemiker Fiedrich Stromeyer, Vorfahren und Seitenverwandte. *Archiv für Sippenforschung, 26*, 130–134.

# References

von Zimmermann, E. A.W. (1783). *Kurze Erklärung der zoologischen Weltcharte*. Leipzig: Johann Beckmann.

Warming, E. (1895). *Plantesamfund – Grundtræk af den økologiske Plantegeografi*. Kjøbenhavn: P.G. Philipsens Forlag.

Willdenow, C. L. (1792). *Grundriss der Kräuterkunde*. Berlin: Haude and Spener.

Willdenow, C. L. (1810). *Grundriss der Kräuterkunde*. Berlin: Haude and Spener.

Willdenow, C. L. (1811). *The principles of botany and of vegetable physiology*. London: William Blackwood.

Withers, C. W. J. (2006). Eighteenth-century geography: Texts, practises, sites. *Progress in Human Geography, 30*, 711–729.

Wittwer, W. C. (1860). *Alexander von Humboldt: Sein wissenschaftliches Leben und Wirken den Freunden der Naturwissenschaften dargestellt*. Leipzig: Weigel.

Wulff, E. V. (1943). *An introduction to historical plant geography*. Waltham: The Chronica Botanica Company.

Zimmermann, E. A. W. (1777). *Specimen zoologiae geographicae, Quadrupedum domicilia et migrationes sistens*. Leiden: Theodorum Haak.

Zimmermann, E. A. W. (1778–1783). *Geographische geschichte des menschen, und der allgemein verbreiteten vierfüssigen thiere*. Leipzig: Weygandschen Buchhandlung.

# Chapter 4
# Classification Divided

> *In the field of plant geography, research has taken especially two courses, a systematical, which is ultimately based on Linnæus's observations and theories in connextion with the distribution of the plant species, and a morphological, which has its origin in Humboldt's theories on the morphological association of different vegetable types with different countries and forms of landscape. The two tendencies have exerted a mutual influence and have, each in its own way, been influenced by the doctrine of descent and its attempt to explain the origin of species out of conditions of geographical distribution (Nordenskiöld 1936, p. 560).*

> *A rift developed, however, between ecological and taxonomic plant geographies – a rift that apparently persists today (Hagen 1986, p. 198).*

Nothing divided eighteenth and nineteenth century naturalists and taxonomists more than the topic of natural classification (see Stevens 1994). The purpose of this chapter is to show the how the practise of natural classification had influenced plant geography, which Nordenskiöld quite rightly states, "has taken especially two courses" (see Nicolson 1996 and below). Nordenskiöld's claim may be used as a road map to show how the practise of natural classification had led to two very different ways of interpreting and classifying the plant geography. The division, as will be shown, has its roots in *practise* (i.e., epistemology), rather than in the development of ideas (i.e., theory), establishing a significant divide between how one classifies between how one interprets the natural world. The emerging roles of classifier and interpreter, within plant and animal geography, in the late eighteenth and early to mid-nineteenth century will be explored below.

## Natural Classification and the "Two Courses" in Plant Geography

Nordenskiöld's split is better understood if we were to ask what eighteenth and nineteenth century naturalists wanted from plant geography.[1] Plant distributions alone didn't produce an explanation as to why organisms were distributed the way they were, rather a classification of these distributions yielded patterns that alone were not easily recognisable. In this sense, a classification of plant distributions was essential in order to understand any further processes. Stromeyer had shown that by adopting Zimmermann's approach, namely classifying animals based on their distributions, to botany, he too could divide the plants into distributions, which would go toward explaining distributional laws. For example, Stromeyer's *Specimen* included a detailed outline of two books:

Book 1. Present range and distribution of vegetables over the world.

  Section I. Range of land and sea vegetables.
  Section II. Range of vegetables in general.
  Section III. A revision of the discussion concerning the main laws governing the geography of the vegetable kingdom, and their comparison with those of the geography of the animal kingdom that are published.

Book 2. Former distribution of vegetables over the world

  Section I. Past and present changes. Vegetable migrations, their causes and consequences.
  Section II. Some thoughts about plants and the ancient world of petrified vegetation that remains unknown (Stromeyer 1800, pp. 6–15).

The classification, divided into present and former distributions of living and fossil plants, provided the foundation for a classification which asked "... where the plants presently live, their differences by reason of their position, extent, parts and general character" (Stromeyer 1800, p. 6). The classification of plant geography proposed by Stromeyer was never adopted, possibly because it was never put into practise. Moreover, neither of Stromeyer's proposed two books was published. Had Stromeyer's proposed a monumental work been completed, would have certainly made plant geography a literature-based pursuit, one that relied on travelogues and floras rather than on field-based observations or measurements, like the approach proposed by Humboldt.

---

[1] Joel Hagen however disagrees. While he sees a divergence, the "distinction between taxonomic plant geography and ecological plant geography was never absolute: it would be historically incorrect to portray them as absolutely divergent" (Hagen 1986, p. 212). I will address Hagen's argument below.

## *Literature-Based and Field-Based Research*

The *Grundriss der Kräuterkunde* by Carl Ludwig Willdenow set the standard for plant classification and geography in the late eighteenth and early nineteenth centuries (Fig. 4.1). The *Grundriss* was in its fourth edition (Willdenow 1805) when Humboldt's *Essai* was published in 1807 and translated into English by 1811. In the section on the "History of Plants", Willdenow spells out the laws of plant distribution:

> By the History of Plants, is to be understood the influence of Climate on Vegetation; the changes which it is probable plants undergo from the revolutions of our globe; their dispersion over its surface; their migrations; and, lastly, the means pursued by Nature for their preservation [...] It is well known that warmth is necessary to vegetation [...] Climate influences the growth, as well as the form, of every vegetable product [...] Plants in their wild state remain pretty constant in their appearance, though they vary sometimes; but these variations are inconsiderable, in comparison of what they undergo when they become objects of culture [...] Besides the manner in which we have said it is probable that plants have been dispersed over the globe [...] The plants of the freshwater are more widely dispersed than those of the land [...] The plants that grow at the bottom of the sea are found in all regions [...] The mountainous or alpine plants are nearly the same on all those chains which had formerly been connected [...] Plants like animals are confounded to certain latitudes [...] From what has been said, it may naturally be inferred that, after so many and such various changes as plants are subjected to, it connote but be difficult to ascertain the exact point from which each has originated (Willdenow 1811, pp. 354–432).

While Willdenow's 'History of Plants' may offer an in-sight into what was known, it does not offer a way to practise plant geography. Willdenow's text offered no classification as to the types of plant distributions (e.g., marine plants, desert plants etc.). Also, no regions were ever proposed, as Willdenow was content to use latitudes and longitudes, but was wary that there were physical and geographical elements to plant distributions (think of mountainous regions in low lying latitudes). The lack of a method, *a way to do* plant geography, makes Willdenow, like Link and Linnaeus before him, merely commentators on the laws governing the distribution of plants.

Botanists during the late eighteenth century focused mostly on physiology, taxonomy and natural classifications. The floras produced by Peter Simon Pallas, Joseph Gmelin, were effectively taxonomic lists, revisions at most, of the taxa found in a particular region. Given that taxonomy can be applied to organisms, it is a way to do classification that involves both the researcher and the objects to be studied. With an application like taxonomy, it is no wonder that late eighteenth and early nineteenth century taxonomists took to classifying plant regions and plant forms (vegetation).

The field-based floras written by Gmelin were works that required dedication and taxonomic experience. Before a geography of plants could ever be considered, a researcher requires an expert knowledge of botany, namely, to be able to identify and describe most plant families. Each flora, like that of *Flora Danica* (Oeder 1761), *Flora Sibirica* (Gmelin 1747), *Histoire des plantes de la Guiane Françoise* (Aublet 1775), *Characteres Generum Plantarum* (Forster and Forster 1776), *Flora*

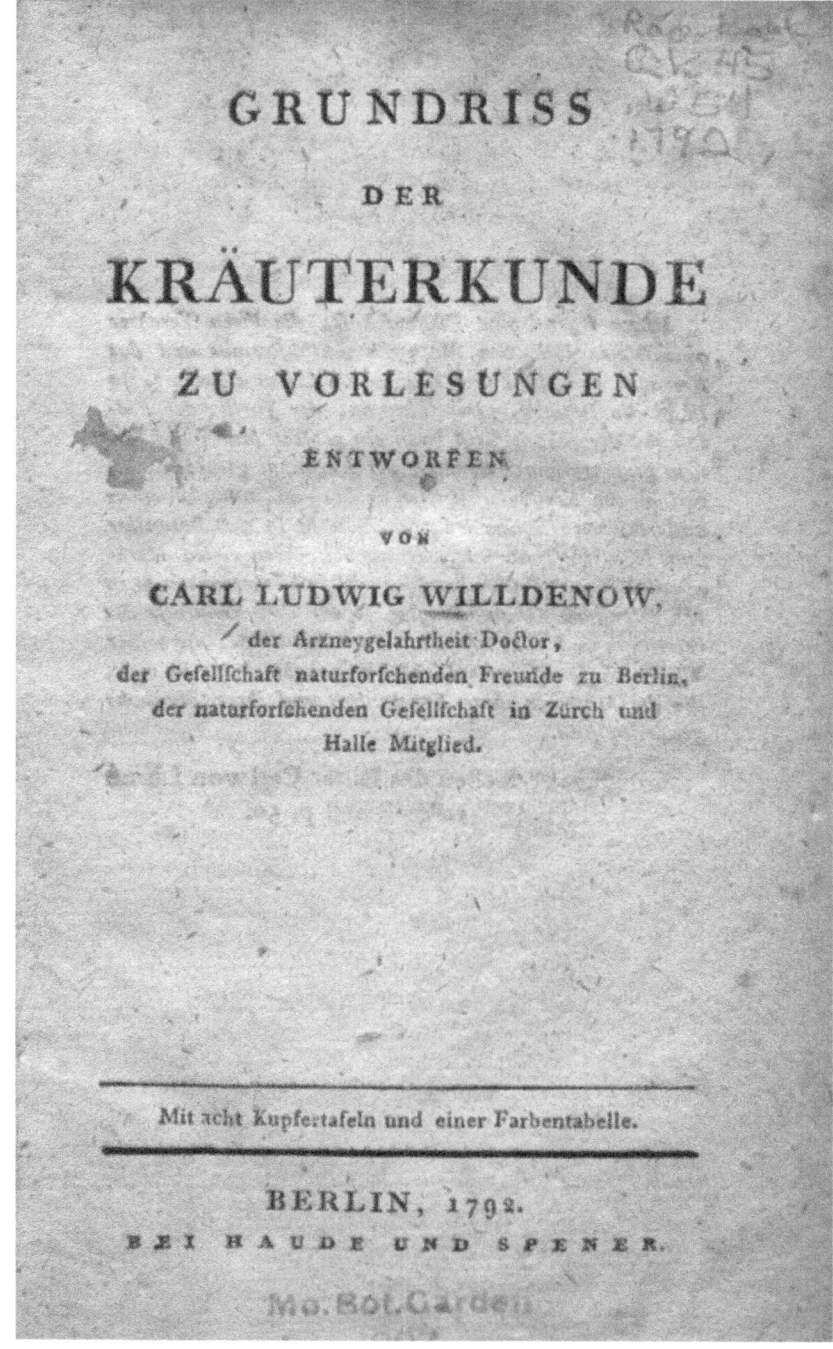

**Fig. 4.1** Frontispiece of the first edition of *Grundriss der Kräuterkunde* by Carl Ludwig Willdenow (1792)

*Pedemontana* (Allione 1785) and, *Flora Rossica* (Pallas 1784–1788), are essentially lists of species descriptions that largely follow the binomial system of Linnaeus. Of the late eighteenth century floras, *Florae Fribergensis Specimen* (Humboldt 1793) stands out in its reference to "zoological geography, whose foundations have been laid by Zimmermann; and the geography of plants, which our colleagues have left untouched" (Humboldt 1793, p. ix, in footnote). Humboldt's *Specimen*, like most eighteenth century floras, was a collection of species descriptions, however, it does include where the plants are found. Humboldt did not create a list of regions nor did he propose a law of distribution. This simple small step and an acknowledgement that plant geography was largely untouched, was perhaps the catalyst for practical approach or method for a field-based geography of plants.

Stromeyer's *Specimen* aside, there was few plant geographies published at the turn of the nineteenth century. Even the Lamarck and de Candolle's 1805 *Flore française* was mostly a literature-based work, which included a geographical map that featured the plant regions of France, with an explanation as to how they were constructed.

> The botanical map of France, which we thought would be a useful addition to this book, is designed to highlight two very different things: (1) the knowledge of vegetation in different parts of France that are known by Botanists; and (2) the general plant distribution on French soil (de Candolle in Ebach and Goujet 2006, p. 766).

*Flore française* differs widely from Zimmermann's *Specimen Zoologicae Geographicae* as it departs from a purely physical geographical approach of gradients and continents to regions defined by the flora and its interaction between soils and climate.

> The map should be considered more of an attempt to apply a specific methodology rather than an attempt to show the complete plant geography of France [...] The area coloured green, marking the coasts from Ostende [Belgium] to Oneille, indicates the realm of maritime [aquatic] plants [...] The blue colour represents areas in France that are occupied by mountain plants [...] The area in vermilion provides us with knowledge of French provinces where the vegetation is intermediate between those of northern plains and those of southern provinces (Lamarck and Candolle in Ebach and Goujet 2006, pp. 766–767).

The map as a tool and the explanation as a guide to deriving maps was possibly the first attempt at a method to classify natural geographical regions,

> I place great importance on altitude as a major factor influencing temperature. I also believe that temperature and not air density greatly influences vegetation, a factor that well known scientists support [...] From the preceding considerations, I believe that in a given country, such as France, the causes that determine the plant region [habitation] could be reduced to three: 1. Temperature [...] 2. The mode of watering [and,], 3. The degree of soil tenacity or mobility (Lamarck and Candolle in Ebach and Goujet 2006, p. 768).

With known plant distributions derived from existing floras, topography (derived from maps) and the three factors that determine plant regions, a botanist could in theory determine the plant regions of France. De Candolle had effectively made plant geography practical and literature-based, meaning a single person could derive a plant geography of a large area like France without having to undertake time consuming and expensive fieldwork. Unfortunately de Candolle's method had little time to shine. Firstly it was a three-volume set that was never translated. Moreover, a general methodology like the one proposed by de Candolle would not have had the wide circulation in a volume that is dedicated to a single area. Whatever the reason, by 1807 Humboldt and Bonpland's *Essai sur la géographie des plantes* had by its very title a wide-ranging appeal.

The *Essai* is divided into two parts, a section outlining the rationale of the geography of plants and the other a methodology, namely *Physical Tableau of Equatorial Regions*. The first section introduces the concept of a vegetation,

> The geography of plants does not merely categorize plants according to the various zones and altitudes where they are found; it does not consider them merely in relation to the conditions of atmospheric pressure, temperature, humidity, and electrical tension in which they live; it can discern just as in animals, two classes having a very different kind of life, and, so to speak, very different habits (Humboldt and Bonpland 2009, p. 65).

Humboldt's geography of plants differs widely from de Candolle (1805) as it attempts to move beyond categorising "plants according to the various zones and altitudes where they are found". Moreover it is unique in proposing the concept of "two classes having a very different kind of life, and, so to speak, very different habits". What Humboldt is referring to are individual or isolated plants and, communities of plants that "live in an organised society like the ants and the bees, and occupy immense terrains from which they exclude any heterogenous plants". Humboldt continues,

> It would be interesting to show on botanical maps the areas where these groupings of similar species of plants live. These maps would show long bands, whose irresistible extension causes the population of states to decrease, the nations to be separated, and creates stronger obstacles to communication than do mountains and seas (Humboldt and Bonpland 2009, p. 65).

Here Humboldt is referring to latitudes and longitudes, as a way to map vegetation. The long bands would reflect latitudinal gradients, which can be easily measured (e.g., what plants are found at a particular latitude, based on air temperature, air pressure etc.). Humboldt's geography of plants was quantitative and introduced a new form of classification,

> Among the variety of plants covering the surface of our planet, one can easily distinguish certain general forms under which the others can be subsumed, and which can be arranged into families or groups that are more or less analogous to each other (Humboldt and Bonpland 2009, p. 73).

Humboldt is referring to *general plant forms*, not taxonomic groups. Forms may be classified into families or groups based "on physiognomy [and] have almost nothing in common with those made by botanists who have hitherto classified them according to very different principles". Humboldt is making a distinction between a classification based on vegetation and the traditional taxonomic approach,

> Only the outlines characterising the aspect of vegetation and the similarities of impressions are used by the person contemplating nature, whereas descriptive botany classifies plants according to the resemblance of their smallest but most essential parts, those relating to fructification (Humboldt and Bonpland 2009, p. 74).

One wonders what Humboldt may mean by "the person contemplating nature"? Can only field-based naturalists see these vegetations, while the literature-based naturalists are limited to looking at herbarium specimens? The *Essai* is a very descriptive and personal account of vegetation and one would not deny that Humboldt has seen the vegetation he describes. But could a naturalist working from the confines of their institution answer the questions that Humboldt poses?

> "Such are the characteristics found in agriculture and its various products, varying with their latitude or with the origin and needs of the peoples. The impact of food [...] the history of navigations and wars" declares Humboldt "are the factors that link geography of plants to the political and intellectual history of mankind [...] These relationships would be no doubt sufficient to show how extended is the science which I am attempting to outline here" (Humboldt and Bonpland 2009, p. 73).

The relationship between man and nature is a common theme in geography, but entirely absent in later works by naturalists who concentrate on plant regions, taxonomy and physiology.[2] But Humboldt was seemingly determined to include the observer into his geography of plants "He will delight in examining what is called the character of vegetation, and the variety of effects it causes in the soul of the observer". Humboldt was interested in the "intellectual cause" of experiences [feelings] that grab our attention. Clearly this derives from Goethe's *Versuch die Metamorphose der Pflanzen zu erklären [Metamorphosis of Plants]*, (Goethe 1790) in which the observer is essentially a part of idea or theory of the objects under study. The questions that Humboldt raises reflect this: "How different is the aspect of a vast prairie surrounded with a few clumps of trees from that of a dense, dark forest of oaks and pines? How different are the forests of temperate zones from those of true equator where naked and thin trunks of palm trees soar above the flowering mahogany trees and resemble majestic porticos? Are they produced by nature, by the large size of these ensembles, by the outline of their shapes, or by the plants' posture?" As Goethe's *Metamorphosis* helped to uncover a basic homology or *urhomology* (Brady 1972; Ebach 2005), Humboldt's vegetation could an attempt to uncover a natural unit of classification within nature. Whatever the reason it did lead to some very interesting questions "What is the character of tropical vegetation? What features distinguish the African plants from these of the New World? What are the analogies in shape that link the alpine plants of the Andes with those of the high peaks of the Pyrenees?" (Humboldt and Bonpland 2009, p. 73).

These questions seem modern and appear to reflect, in part, those of Buffon. Humboldt was after something more than a classification or law. He was after the individual factors that shape nature, the hidden processes, wherever they may be. Moreover, to Humboldt's mind, a vegetation is the result of these processes and therefore is a unit of a natural classification. After all, if factors like temperature, altitude and geology, do create "groupings of similar species", then surely Humboldt

---

[2] With the exception of distributions directly or indirectly influenced by humans.

must have been convinced he had stumbled upon natural forms. Moreover, Humboldt could *measure* these factors and *record* the types of vegetation. The geography of plants was far more than a list of laws, like those listed in Willdenow and far more than mere observation. Humboldt had combined measurements and observation with known processes and, technology, into a single method. The method did not focus on one topic, such as taxonomy, rather it incorporated as much information and knowledge about a region into a single useable methodology, which Humboldt outlined in the *Essai* as the *Physical Tableau of Equatorial Regions*.

The *Tableau* includes plant and animal distribution, geology, temperature, snow lines, barometric pressure and the boiling point of water. Together these measurements seem excessive in order to determine a geography of plants. However Humboldt wanted it all – every possible influence on plant distribution, physiology and behaviour. The *Tableau* was more a physical geography of everything rather than a geography of plants. The work, "based on measurements and observations performed on location, from the tenth degree of boreal latitude to the tenth degree of austral latitude in the years 1799, 1800, 1801, 1802, and 1803" (Humboldt and Bonpland 2009, p. 76). Perhaps this is why Humboldt's approach was never fully adopted. Who at the time had the resources to collect that amount of data over a 5-year period?[3] The appeal the *Tableau* did have is that it was achievable (given the cost) and, represented real observations and measurements on a map that could be used and understood by a wide range of naturalists.

In 1813 Johann Wolfgang von Goethe published a systematic comparison of the heights of the new and old worlds also as a tableau titled *Heights of the Old and New World, figuratively compared* (Goethe 1813, Figs. 4.2 and 4.3):

> In 1807 our most excellent Alexander von Humboldt sent me his Geography of Plants [...] I devoured the work and wanted it to be fully enjoyed and made useful to others, however I have been prevented somewhat as my copy lacked a map (Goethe 1813, p. 5).

Goethe adopted Humboldt's mapping technique directly from the text and drew his own tableau (Goethe 1813), rather than adopting Humboldt's plant geography overall. As Stephens correctly points out, Humboldt's *Tableau* as it is depicted in his *Essai* is not systematic, that is, allowing it to be compared to another region (as in de Candolle's method). "Goethe filled the breach in parts; after reading a draft of Humboldt's *Essay*, Goethe drafted a profile comparing the 'ancient continent' with the 'new continent'" (Jackson 2009, p. 29, footnote 46, original emphasis). Goethe's systematic arrangement however did lack the detail of Humboldt's *Tableau*. While Humboldt's map is based on a vast amount of measurements, Goethe seemingly based his on what was available in the literature mostly latitude, altitude and snow

---

[3]There were some exceptions. For example, Franz J.F. Meyen did conduct a large expedition between 1830 and 1832, however these were nowhere near the immensity of Humboldt's expedition (see Nicholson 1996, p. 294).

**Fig. 4.2** Goethe's *Heights of the Old and New World, figuratively compared* (1811). The full title of the map reads *Höhen der alten und neuen Welt, bildlich verglichen* (Source: Kartenarchiv Plus, courtesy of Dr. Andreas Christoph, Friedrich-Schiller-Universität Jena. http://www.leopoldina.org/de/ueber-uns/studienzentrum/projekte/kartenarchiv-plus/)

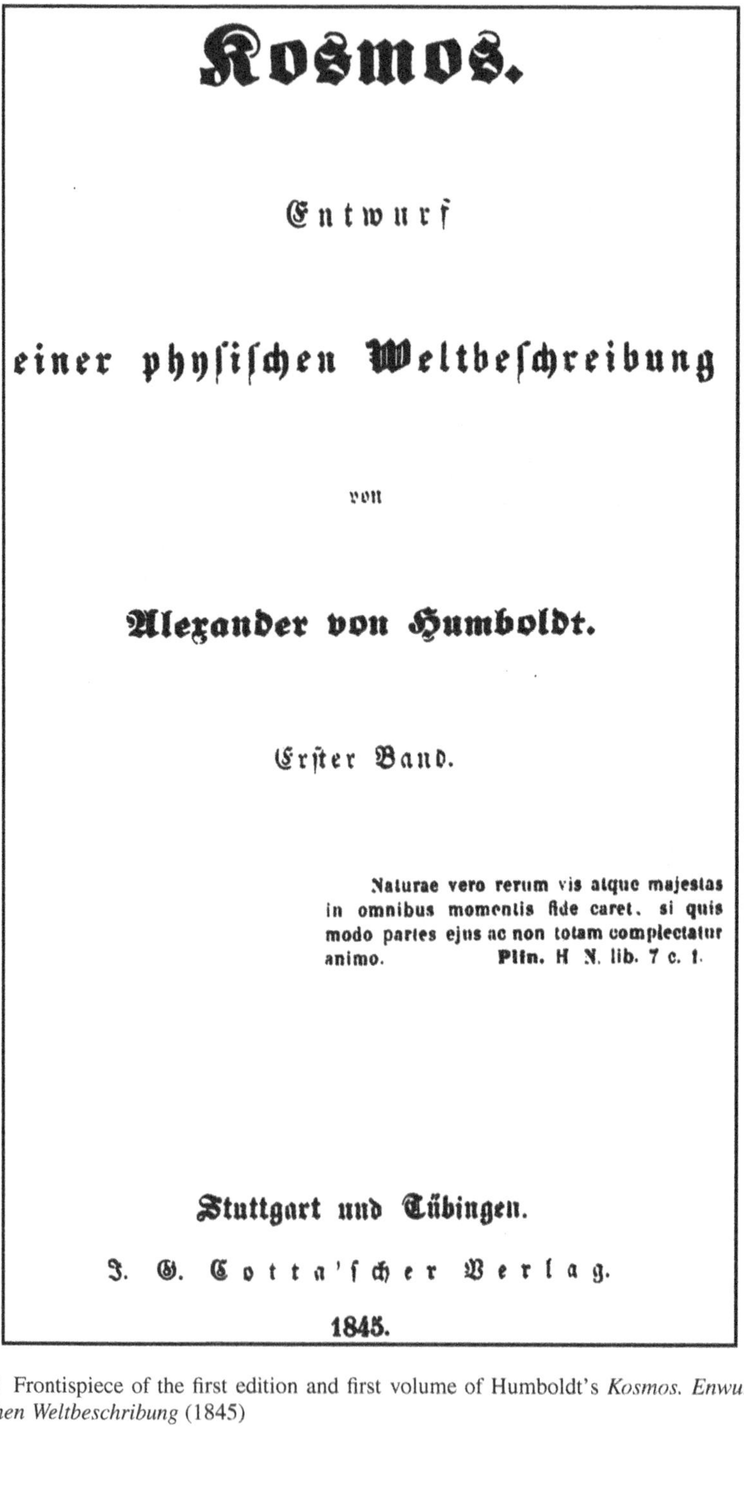

# Kosmos.

## Entwurf

## einer physischen Weltbeschreibung

von

### Alexander von Humboldt.

Erster Band.

> Naturae vero rerum vis atque majestas in omnibus momentis fide caret, si quis modo partes ejus ac non totam complectatur animo. — Plin. H. N. lib. 7 c. 1.

Stuttgart und Tübingen.

J. G. Cotta'scher Verlag.

1845.

Fig. 4.3 Frontispiece of the first edition and first volume of Humboldt's *Kosmos. Enwurf einer physischen Weltbeschribung* (1845)

lines.[4] Regardless, Goethe had managed to implement a method or technique that compared the heights of two continents.

Lateral mapping techniques (or phytogeographical profiles) like that of Humboldt and Caldas were not adopted in plant or animal geography. Humboldt's legacy was not his cosmological geography, one that mapped heights, distributions of plants and the elevation of "small fleecy clouds" up mountain sides, or even the collection of casts amounts of data, ranging from temperature to "the chemical composition of the atmosphere [and] its electrical tension". Humboldt's legacy resides in his practise of classifying plants as vegetation types, (based on physiognomy) rather than as taxa, something that Nicolson rightly claims:

> The study of plant physiognomy was an important feature of Humboldt's botanical enterprise and this aspect of his investigation of vegetation constitutes one of the most decisive ways in which he departed from Linnaean taxonomic methods. Classification by life-form, although in some cases it might approximate to more orthodox arrangements, was essentially independent of floristic systems [...] Similarly, species closely allied by the taxonomist might be very different physiognomically. One of Humboldt's reasons for according a central importance to plant geography within the scheme for a universal science was his conviction that the vegetation of any given region was not only a primary expression of the physical environment; it also exercised a formative influence on humanity, both materially and spiritually (Nicolson 1996, pp. 291–292).

In departing from Linnaean taxonomy Humboldt had set up a precedent; namely the establishment of a new science, one that does not take taxic distributions into account. In doing so, Humboldt had divided plant geography between the taxonomists who studied plants as taxa and the geographers who studied plants as vegetation.

## *Humboldt's Legacy and the Classification of Vegetation*

The development of Humboldt's plant geography in the nineteenth century is covered by the Historian of Science Malcolm A. Nicolson in his insightful paper *Humboldtian plant geography after Humboldt: the link to ecology* (Nicolson 1996).[5] Nicolson tells how the split occurred in regards to Humboldt's methodology being adopted by plant geographers such as Joachim Frederik Schouw, Franz Julius Ferdinand Meyen, Anton Kerner von Marilaun and, August Heinrich Rudolf Grisebach (see below).

---

[4]Goethe's 1813 map also depicts "v. Humboldt" on the slope of Mount Chimborazo who appears to be waving to a certain "de Saussure" on top of Mont Blanc.

[5]Nicolson (1996) presents the best history on the division between the floristics of taxonomists and the vegetational classification of early plant geographers (Humboldtians). Unfortunately the paper is never cited by scientists, particularly those who have commented on the history of ecology and biogeography (e.g., Jax 2011). I base this claim on the 23 citations for Nicolson (1996) listed in Google Scholar (accessed January, 21, 2014).

Nicolson's Humboldtians adopted of certain aspects of Humboldt's geography of plants. Climate was adopted as the main driver for plant distribution (including altitude). Plant forms (i.e., vegetation), were also adopted in favour of species (which were used, as a key to identification). Maps were also used, but as normal Mercator-style projections, rather than that of Humboldt's *Tableau*. In doing so, the Humboldtians had circumvented the holistic approach of Humboldt. While Humboldt sought to look at very aspect of plant distribution, his cosmological approach was perhaps to overwhelming for those tied to their desks and unable to venture out and collect the vast amounts of data. The first practitioner of Humboldt's method, Danish naturalist Fredrick Schouw, attempted to make it practical. In his *Grundzüge einer allgemeinen Pflanzengeographie* (Schouw 1823), taxonomic groups (i.e., families and genera), such as "the palms", were used to describe broad distributions. Schouw termed this *Örtliche Verhältnisse* ("Regional relationships"), while vegetation types, used to describe distributions within smaller areas, was given the term *Botanische Geographie order Vergleichung der verschiedenen Erdteile in hinsicht ihres vegetative Erzeugnisse* ("Botanical geography or the comparison of the different parts of the world in respect to their vegetative products"). For Schouw and the Humboldtians, vegetation and the climatic factors governing its distribution was plant geography. But this did not rule out traditional taxonomic practise. Schouw still relied on species names and their distributions to classify vegetation. For example, coastal vegetation's were described by the physiognomy, such as small leaves, however, species and genus names were still used to identify of what the vegetation consisted. Given this, early plant geographies were heavily reliant on existing floras, their distributions and climatic measurements. Regardless Schouw's plant geography was a break from traditional botanical practise:

> Willdenow is, so far as I am aware, the first to attempt a comparison of floristic regions; without a doubt, it lacked present day distributions and relied on arbitrary historical plant hypothesis (Schouw 1823, p. 502, my translation).

Schouw also criticised de Candolle's use of 20 floristic regions (de Candolle 1820):

> Candolle compares 20 floras, or as he calls them, regions. In his method, which he has developed studying these floras, [Candolle] does not reveal the characteristics that each form takes; it appears that the main basis for the division [of the regions] is current distributions (Schouw 1823, p. 504, my translation).

Not convinced of de Candolle's method, Schouw provides a set of rules to delimit plant geographical regions:

> 1. at least half the known species are particular [endemic] to a region; 2. that 1/4 of the genera are either fully [endemic] or mostly occur in a region. 3. That single families are either fully [endemic] or mostly occur in a region (Schouw 1823, p. 504, my translation).

Compare Schouw's rules to those of de Candolle (1820):

> In his section on habitations, Candolle makes the following points: (1) every species tends to occupy a certain space, and the determination of the laws that govern species distributions is the study of habitations; as (2) there are more species in the tropics than in high latitudes;

36 (3) the numbers of species of monocots and dicots vary in certain ways; 37 (4) certain numbers of species are recorded for certain countries (de Candolle 1820, pp. 392–400 in Nelson 1978, p. 281).

Schouw's list of rules is far more precise than de Candolle's.[6] Schouw proposed 22 regions (compared to de Candolle's 20), which he named based "on plant form characteristics" rather than on common geographical names proposed by Willdenow, Treviranus and de Candolle. The mixture of vegetative and floristic elements in Schouw's regions demonstrates the eclectic nature of early nineteenth century plant geography. Taxonomists like de Candolle relied on a descriptive classification, one without a complex methodology. Although Candolle attempted to describe a method based on altitude and temperature (see below), it was the transition from a taxonomy of species and genera to vegetation types that secured the success of the Humboldtian method.

While vegetation types were seemingly arbitrary (e.g., descriptive rather than prescriptive), they could be quantified through physical measurement (e.g., altitude, temperature, barometric pressure etc.) and described via their physiology. The practical implications were immediate: expert knowledge in taxonomy was limited. Limited knowledge also restricts ones practical abilities. Taxonomy is more than the identification and description of new species and the seemingly "legalese" application of nomenclature. Taxonomy is mostly about revision, improving existing taxonomies based on the evidence of homologs, the parts assigned as evidence to support taxonomic groups (see Wilkins and Ebach 2014). Naturalists, who lacked a working knowledge of taxonomy, may have created artificial classifications within plant geography. The claim I am making here is that plant geography was available to non-taxonomists and to non-specialists. Anyone could do it. For example, the influential text *Grundriss der Pflanzengeographie* (Meyen 1836), which appeared 13 years later, was an attempt at a classification of plant forms, vegetation and climatic zones, however the author, Franz J. F. Meyen, a Prussian plant anatomist, physiologist and phyto-geographer, did not practise taxonomy.

Meyen's *Grundriss* "represents one of the earliest, most successful and most explicit articulations of the Humboldtian exemplar [and] was, among other things, a sustained and sophisticated attempt to correlate vegetation with measured physical factors" (Nicolson 1996, p. 294). Nicolson describes Meyen as having an interest in systematics (the study of classification). However, he had not shown this in practise, as he preferred a classification of physiognomy.

Meyen's *Grundriss* concerns itself with plant physiognomy and how climate (i.e., weather) affects the distribution of plants:

---

[6]Nelson summarises the two methods concisely: "A point of confusion in recognizing areas of endemism was created early in the history of plant geography by J. F. Schouw, *Grundzüge einer Allgemeine Pflanzengeographie* [...] To qualify as botanical regions according to Schouw, areas must exhibit certain qualities beyond the elder Candolle's concept" (Nelson 1978, p. 297, footnote 91).

It is very easy to show that conditions of climate particularly heat and moisture, are the chief causes which determine the station and distribution of plants; and therefore it is of great importance to the science of botanical geography, to know exactly the modes in which the influence of the often extremely complicated conditions of climate becomes apparent (Meyen 1846, p. 8).

Climate (more correctly weather and mean temperature), marked out the areas of vegetation:

... therefore, we may now pass on to the application of the mean temperature to the purpose of botanical geography [...] Baron Alexander von Humboldt has here also marked out the path of this science. He connected those places which possess an equal degree of heat by lines, which he called isothermals (Meyen 1846, p. 20).

When climate was unable to explain certain distributions, Meyen turned to soils:

... we now go on to the consideration of the many local conditions which can promote or hinder the presence and distribution of plants, even though heat and moisture be present in sufficient quantity. These conditions are, for the most part, the relations in which plants stand to the soil in which they grow, and the consideration of them is one of the principle objects of botanical geography (Meyen 1846, p. 46).

Areas of vegetation were divided into longitudinal zones, which were "marked out by an eastern and western limit. The regions of plants, or their vertical range, are marked by an upper and an under limit, which are defined by degrees of altitude. The area of a plant, or its range, is either uninterrupted or interrupted" (Meyen 1846, p. 89). Together this constituted a distribution of plants, which was governed by distributional laws, mainly driven by climate and partially by soils. "The subject of the distribution of plants" Meyen continues "may be divided into two perfectly distinct branches [...] Physiognomics considers vegetations according to the distribution of the forms which point out the groups of plants". Meyen considered this to be a natural system in which the "similarity of form is the principle of classification". The key to physiognomics of vegetation was to "investigate the predominance" of plants forms based on the "absolute mass of its individuals". The statistics of plants, however, did not concern itself with the predominance of plants groups, rather it looked at the "relative proportions, founded on real numbers, which this or that group by its number of species bears either to the whole mass of known plants, or to the number of species of other groups". For example, when a single widely distributed species of fern may cover the same area of an entire family of plants (e.g., Compositae), one species of fern will dominate by the, mass of individuals rather than the number of species. The result is a statistical sum "the Compositae and the ferns stand in the proportions of 1.13 and 1.15 to the sum of all the phaenogamous plants" (Meyen 1846, pp. 98–99).

Although a division exists in theory, between taxonomy and physiognomy, it was not obvious in practise. Species and families were still used to distinguish between individuals. Even at the level of a classification of areas, Meyen sees two separate parts to a single classification – the geographical and true botanical – in which the physiognomy, sometimes agrees with the artificial [taxonomic] characters, as in Schouw's regions. If the geographical were taken as the principle foundation, the world would be divided into floras based on countries, similar to the regions of de Candolle (1820) and Schouw (1823). Furthermore, the countries could be divided into vegetation and, the vegetation into forms. Meyen's non-hierarchical classification was driven by temperature in latitude and in altitude and delineated by isothermal lines. For example, the 'Fern Form' "... may be classed in three division, viz: the herbaceous, the shrubby, and the arborescent. The herbaceous ferns chiefly belong to the temperate and cold zones [...] but we choose the name [Shrubby Fern] in order to distinguish them clearly from the tall, arborescent Ferns" (Meyen 1846, p. 126). Meyen still uses species names to distinguish between the different types of ferns. The horizontal or latitudinal division of the globe includes for instance the 'Tropical Zone' "... stretches on each side of the equator from the 15th degree of latitude to the tropics, and shows a mean heat of from 23° to 27° Cels." (Meyen 1846, p. 199). The altitudinal division includes the 'Region of the Tree-Ferns and Figs' "... commencing at the height of 1900 feet, this region stretches as high as 3600 and 3800 feet, possessing mean temperature of from 22° to 23.5° Cels." (Meyen 1846, p. 231). Meyen's classification does not include Schouw's or de Candolle's geographical regions. Rather Meyen has a precise delineation of each area (by isothermal lines), as opposed to de Candolle's regions.

Meyen's approach however appears to be arbitrary. Fern Forms are vague at best and are clearly intended to replace species, genera or families, the units used to delimit de Candolle's regions. Moreover, the Region of the Tree-Ferns and Figs does not overlap with the Tropical Zone, as ferns are found in other Zones (e.g., Tropical Zone, Sub-Tropical Zone, Warmer Temperate Zone). A partially overlapping classification is artificial, as Form, Vegetation or Zone would not constitute a group in their own right. For example, while there is only one Tropical Zone, there would be multiple instances of the 'Region of the Tree-Ferns and Figs', neither of which are related in species composition or by the climatic factors that shape them. In this sense each instance would be a new and distinct region. One would benefit in viewing Meyen's classification as two separate, partially overlapping, classifications, that together form an artificial classification that is in stark contrast to de Candolle's floristic regions and sub-regions. De Candolle's classification is based on a far more stable taxonomy of species, genera and families. Regardless, Meyen considers de Candolle's areas as "physiognomic" as they seemingly resemble Schouw's vegetations. Meyen continues:

> To have an exact acquaintance with these principal forms of vegetation is of the greatest importance to a phyto-geographical division of the globe, as they principally fix the natural physiognomy of different countries. Humboldt is the first who has made such a classification of vegetation, and this must be taken as the foundation of all further inquiry into the subject. It is not until we are somewhat intimately acquainted with the various characteristic forms of plants, that we shall be able to recognise the peculiarities of each flora, and to characterise the physiognomy of each country (Meyen 1846, p. 106).

Meyen claim appears to state that his method subsumes all forms of plant geographical practise. Taxonomic characters are presumed to be "artificial", when compared to the vague forms proposed my Meyen. Furthermore assuming a "natural physiognomy" that is driven by climate, one would expect significant

**Table 4.1** Hierarchical units of physiognomic classification (1830s–1870s) taken from Whittaker (1962)

| Author | Hierarchical units of physiognomic classification (1830s–1870s) | | |
|---|---|---|---|
| Heer (1835) | Regions | Localities* | Forms |
| Meyen (1846) | Zones | Regions | Forms* |
| Sendtner (1854) | – | Vegetation–forms* | Sub–forms |
| Lorenz (1863) | Regions | Facies* | – |
| von Marilaun (1863) | Regions | Vegetation–formations* | Species–formations |
| Grisebach (1872) | Formation* | – | – |

After Grisebach (1872), Formations were used to signify the basic unit within ecological plant geography. Names denoted with an asterix (*) are considered by the respective authors as the functional unit with plant geography

overlap between Forms and Zones, which is lacking in Meyen's classification. Most astonishing, is that the assumption of a natural physiognomy over an artificial taxonomy somehow subsumes de Candolle's plant classification into Schouw's vegetational classification. Why then, was Meyen so eager to dismiss taxonomic classification as artificial? The most likely explanation is that units of vegetation can be explained by abiotic factors, like temperature, rainfall, soils and so on, where as, the nature of a species, was and still is, a hotly debated topic (see Wilkins 2009). However, Meyen's forms belong to a largely electric mix of vegetation units (see Whittaker 1962). Given that vegetative plant geographers were keen to establish a natural unit of classification, each author seemingly had devised their own. By the time Grisebach published his *Die Vegetation der Erde nach Ihrer Klimatischen Anordnung* (Grisebach 1872), there were fewer than five classification systems and natural units of vegetation (Table 4.1). Writing in 1910, C. E. Moss observed that "the subject of ecological plant geography has suffered and still suffers very considerably from a lack of uniformity in the use of its principal terms" (Moss 1910, p. 18), a sentiment echoed by David W. Shimwell 61 years later "the subject of vegetation description and classification is extremely diverse and complex, and in no other branch of biology or indeed science does the Latin maxim *quot homines tot sententiae* apply more aptly than to the diversity of opinions expressed on the subject" (Shimwell 1971, p. xiii).[7] For a field seeking a natural classification in vegetation rather than in taxonomy, ecological plant taxonomy had suffered from a multiplicity of classifications in an age when natural classification was becoming firmly established in taxonomy.

---

[7]Hagen (1986, p. 210, footnote 37) notes Shimwell's concerns further "The classification of vegetation suffers greatly from overstatement, some ambiguity and, inevitably, misinterpretation. The history of vegetation classification is chaotic" (Shimwell 1971, p. 42). See also Stott (1984) who in his *History of Biogeography*, claims that there "were two major objectives. The first was the description and classification of the world's vegetation formations". The second was "the discovery of 'ecological units' of nature, in contradistinction to the systematic or taxonomic units of the taxonomist" (Stott 1984, p. 5).

## Artificial and Natural Classification: A.P. Candolle and His Geography of Plants

Meyen's *Outlines of the Geography of Plants* was first published in German in 1836, at a time when plant taxonomy had established a natural classification, mainly through the work of Antonine Jussieu and Augustin Pyramus (herein A.P.) de Candolle (de Candolle 1813; Williams and Ebach 2008).

A.P. de Candolle's botanical geography was radically different from that of the Humboldtians. Firstly it did not rely on a classification of vegetation and, on only one geographical factor:

> One of the most important points in botanical geography is to analyse with accuracy the influence which the absolute height of a place above the level of the sea produces upon vegetation. It is a complicated circumstance, depending upon a variety of causes, which are not necessarily connected together; in order therefore, to understand it in all its relations, it is necessary to examine separately all the external agents of vegetation. Height may act upon vegetables either *mediately* or *immediately*; height influences the temperature of the atmosphere and its humidity, and also the intensity of the solar light; but temperature, moisture, and light, all affect vegetation; therefore, in this way, height will act *immediately* on vegetables (de Candolle 1818, pp. 408–409, original emphasis).

A.P. de Candolle had cleverly acknowledged all the possible influences on plants and in doing so, dismisses them by justifying that height (altitude) alone is the only constant to measure against. Note also that de Candolle had abandoned "the mode of watering" and "the degree of soil tenacity or mobility" (de Candolle 1805). De Candolle lays out 6 "laws or general principles respecting botanical geography. 1. the degree of rarity of atmospherical air [ ... ] does not appear to have any very essential direct action upon the geography of plants. 2. The geography of plants of different regions is principally determined by the mean temperature, and by its annual phases. 3. [ ... ] the mean temperature of a given place is determined by the latitude [..] 4. The annual phases of temperature [ ... ] establish a strong relation between the vegetation of very elevated districts and that of the northern countries. 5. Annual and biennial plants [ ... ] become more rare in proportion as we remove from the equator, or from the level of the sea" (de Candolle 1818, pp. 412–413).

A.P. de Candolle was being practical. No single person could measure all the factors listed by Humboldt, Schouw or Meyen, without having to spend time and money to get there. Topography, however, was another matter. It usually was among the first measurements made by travellers and explorers, and if these can be easily accessed in floras or travelogues, then de Candolle's justification makes sense. Why investigate *everything* when a basic unit like height can predict this for you? But de Candolle's method was restricted to much smaller areas, in which topography was influenced by present day climate. It did not explain the differences for example in species between continents. In addition, de Candolle never abandoned taxonomy in favour of adopting a "natural" vegetative classification. The third difference between de Candolle's botanical geography (i.e., taxonomic geography) and the Humboldtian vegetation geography was in how areas were determined. For de Candolle the botanical geography consisted of habitations (regions) and the smaller stations (habitats), while for the Humboldtians the division was between those that investigated the distribution of species, genera and families versus those that looked at the distribution of vegetation types, or forms.

A.P. de Candolle's habitations and stations[8] were used by Nelson (1978) to justify the split between taxonomic geography and vegetation geography. However, this has not been necessarily universal within the history of science (see Nordenskiöld 1936, Hagen 1986, Nicolson 1996). While the size of an area may be resultant of a split between a taxonomic (habitations) and vegetative (stations) classification, it is not historically correct to state that it drove Humboldt's plant geography. Clearly the rejection by the Humboldtians of an "artificial" taxonomic classification, in favour of a natural vegetative unit, is clearly what created the break that resulted in taxonomic and vegetation geography. It may be said that taxonomic distributions fall within regions (habitations), while vegetations form smaller habitations. While this may be a general trend it may not necessarily be the rule. Regardless, de Candolle's stations and habitations is seemingly still used to justify the break between taxonomy and ecological geography in both botany and zoology (e.g., Nelson 1978, Drouin in Acot 1998, Stevens 1994).[9]

Even though this historical split was used to justify two plant geographies, only one established itself with botany. De Candolle's method, seemingly the only taxonomic plant geography, did not have a lasting legacy. After his death in 1841, it was abandoned. His son, Alphonse, had already adopted the statistically inclined method proposed by Schouw (de Candolle 1830 *with* attribution, *contra* Nelson 1978, p. 297, footnote 91) and, had reappraised his father's botanical geography:

> I would divide the globe by region, as proposed so far, for largely artificial systems. The rules are too arbitrary, and the obtained regions are not similar in the majority of books or recognised by the consent of the greater number of botanists (de Candolle 1855, pp. 1304–1305).

Alphonse Candolle (herein A. Candolle) recognised there were problems with regionalisation, given that different botanists had a different set of regions. But A. de Candolle was referring to plant geography *as a whole*. The regions of Schouw and A. P. de Candolle may have overlapped, but they were built upon different premises. Moreover, Schouw saw vegetation forms and current climatic processes as the primary basis for a plant geography, while de Candolle clearly saw regions, based

---

[8]The terms stations and habitations have been in use in the eighteenth century. Richard Bradley (1688–1732) in his *Philosophical account of the works of Nature* uses the term station, as the position of an object and, habitations, although he does not define them. The former is used to describe the place in which a organism lives and grows "Of these are the *oyster,* the *muscle* [sic], the *Cockle,* the *Barnicle* [sic], &c. which are never capable of removing themselves from their first station" (Bradley 1721, p. 49). However, Bradley uses the term habitation in place of station: "The Habitation of the Lobster is in Holes among the Rocks, where the Sea never leaves them" (Bradley 1721, p. 53).

[9]Nelson may have derived this from reading Lyell's interpretation of de Candolle (1820) (Lyell 1833), Egerton (1968) and Kinch (1974 Master thesis cited in Nelson 1978).

on taxonomic units (e.g., species, genera) and historical processes (e.g., mountain building), as the more important. A. de Candolle's *Géographie Botanique raisonnée* (Candolle 1855) however changed the emphasis from regions to species.

By rejecting arbitrary regions, A. de Candolle (1855) had "raised geographical botany to its proper rank among the his/her branches of physical science" (Anonymous 1856, p. 492). A. de Candolle did keep the idea of stations and habitations of his father, which he termed botanical geography and geographical botany respectively. Botanical geography looks at "the peculiarities of the of the vegetation of a given country, the relative proportions of the families, genera and species" while geographical botany "examines the distribution of species, genera and families over the surface of the globe" (Anonymous 1856, p. 493). Ironically, A. de Candolle has enforced his father's stations and habitations onto a field that is divided between rival "natural" vegetation and taxonomic classifications. The one difference was Alphonse's rejection of regions, which ironically, is what the elder de Candolle saw as the natural historical unit within plant geography, something that connected a natural classification and distribution with historical causation. In 1855, however, the emphasis on species and their possible common descent (*sensu* evolution) removed the need for regions altogether. Geographical botany (a.k.a. taxonomic geography) and its emphasis on historical "evolutionary" processes presents a modern plant geography, one that has shifted away from natural regions towards investigating the mechanisms involved in natural taxonomic groups, particularly species. At the same time, the Humboldtians (vegetation geography) were looking for natural units elsewhere, in vegetative forms.

## Toward a Unification

*Geographisches Jahrbuch* (1866) *as an exemplar of the divide in animal and plant geography*

> The study of stations has been styled the topography, that of habitations the geography, of botany. The terms this defined, express each a distinct class of ideas, which have been confounded together, and which are equally applicable in zoology (Lyell 1833, p. 72).
>
> ... and next in order of interest, at all events to the naturalist, we should place the essays of Schmarda and Grisebach (G.E.D[10] 1871, p. 44).

In the same 1866 issue of *Geographisches Jahrbuch* Grisebach and Schmarda provide two mildly conflicting accounts of organismal geography, which set the scene for later nineteenth century plant and animal geographical practise. Consider these two remarks:

> One of the most important elements "... is the portrayal of unique characteristics in the natural flora, since arbitrary classifications of the plants contain negative characters, rather

---

[10]G.E.D. may refer to English physician George Edward Day (1815–1872) of Furzwell House, Torquay, Devon, who had contributed several reviews and notes to *Nature* between 1870 and 1871.

only those vegetative formations deserve to be recognised as independent plant forms, which conform to the influence of climate" (Grisebach 1866, p. 384).[11]

On of the basic premises on which the natural sciences thrive is the persistent revision of all its sub-disciplines. Animal geography will also contribute to the development of geography will be instrumental in extending geography and systematic zoology through adjusting the definition of species (Schmarda 1866, pp. 426–427).

In the same volume Grisebach and Schmarda state two very different geographies. Grisebach sees the interaction between physical characters and climate as being natural and, taxonomy as arbitrary. At the same time Schmarda views species as real (i.e., natural) and climate as arbitrary. Hence, Schmarda's zoogeography was centred around taxa (i.e., species) and their endemic areas (Fig. 4.4), while plant geography had centred on vegetation and the how different climates created various types of vegetation. The conceptual split between taxonomy and vegetative classification heralds the division in plant and animal geography.

Schmarda and Grisebach were both unanimous on what caused diversity, namely the "influence of climate on the formation of vegetation causes physiological changes in organic life" (Grisebach 1866, p. 379), they differed about what the natural units were. Moreover, areas, large or small, did not play a vital role as Schmarda notes,

The dependence of animals on their surrounding environment and climate is so great that every geographical region contains its own endemic fauna. To understand these areas is the goal of geographical distribution. Fauna bounded by hemispheres, climatic zones and geopolitical borders have a limited value because they either unite heterogenous regions or they tear up existing larger geographical wholes. Local and regional fauna have their own value when we look at the relationship between these localities, which contain the essential characters of the large region that they represent. As similar geographical areas combine into a large physical complex, for instance the Mediterranean and the Asian Highlands, are such singular fauna that are united within a larger whole, namely a zoological region (Schmarda 1866, pp. 424–425).

While there may be a dispute as to how the Earth is craved up into regions, areas were rarely seen as natural units in the same way that species or vegetable formations were natural. Here I return to the claim made by Nelson (1978) that "the terms as used by Candolle, have modern counterparts: ecological and historical biogeography. Ecological biogeography is the study of stations; historical biogeography, the study of habitations" (Nelson 1978, p. 281).[12] Clearly this is

---

[11]Greisbach refers to a text titled "Über die Grenzbestimmung der Vegetationsgebiete, die geobotanische Einteilung der Erde, s. 'Geogr. Mittheil.' 1866, Heft II, mit Karte" (Greisbach 1866, footnote 1, p. 407, translated as On the definition of the boundaries of areas of vegetation, the geobotanical classification [division] of the Earth, see 'Geogr. Mittheil.' 1866, Volume II, with map). The citation refers to Grisebach, A. (1866). Die Vegetations-Gebiete der Erde, übersichtlich zusammengestellt. Mittheilungen aus Justus Perthes' Geographischer Anstalt 12: 44–53.

[12]It is worth pointing out that in her book *Modern Nature* Nyhart (2009) claims that the split between ecological and historical biogeography "first crystallized in the consciousness of ecological animal geographers" (Nyhart 2009, p. 324). I believe that this is historically incorrect. Perhaps Nyhart misunderstands what ecological and historical biogeography mean in the context

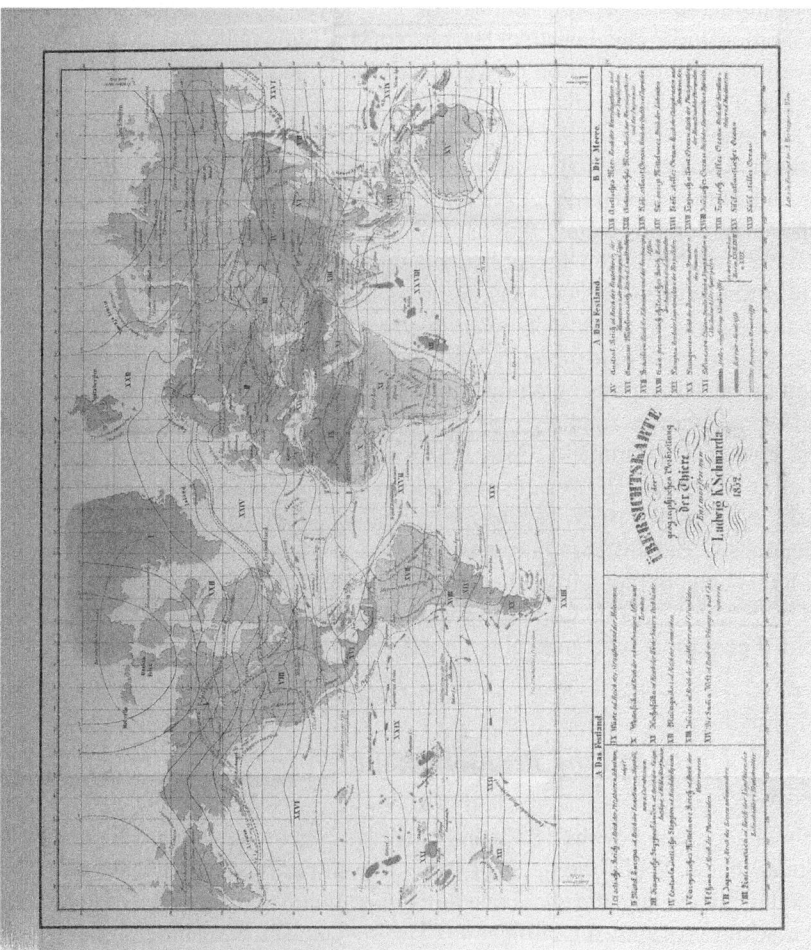

**Fig. 4.4** *Übersichtskarte der geographischen Verbreitung der Thiere. Entworfen von Ludwig K. Schmarda. 1852* [translated as: A summary map of the geographical distribution of animals]. The map is located in the back of the third volume of *Die geographischen Verbreitung der Thiere* (Schmarda 1853). The map contains the 21 terrestrial regions and 10 marine regions that are outlined in all three volumes of Schmarda (1853). The distributional limits of vertebrate genera and/or families are drawn in dotted lines

not necessarily the case as demonstrated above. Schmarda represents biological taxonomy and the distribution of species of taxa across the globe. Although Schmarda worked primarily on vertebrates, his view that species are natural units with a natural taxonomic classification of genera and families conforms to that of Augustin and Alphonse de Candolle. Grisebach like Nicolson's Humboldtians were dismissive of a natural taxic classification of species genera and families. They saw plant forms, not species, as natural units and biological classification as an arbitrary agglomerations of "contain negative characters". Given that historians such as Erik Nordenskiöld and, later Joel Hagen and Malcolm Nicholson have often noted the divergence in classification between early plant and animal geographers, why suggest that it was otherwise?

I wish to keep the division between plant and animal geographers and its history separate from the history of regionalisation, which I will cover in the next chapter. While plant and animal geographers could be divided between those who saw organismal forms as natural (mainly those studying plants) from those who viewed species as natural (mostly those who studied animals), both fields used stations and habitations to differentiate between local and regional geographies (see Chap. 5). Rather, it was how these areas were *identified and diagnosed* which differed. For example, Grisebach would identify both floral (habitations) and vegetative (stations) regions, based on physiognomic groups (i.e., vegetation), while Schmarda would identify local (stations) and regional (habitations) faunas based on taxonomy (i.e., species, genera and families). Seen this way, the historical and ecological biogeographers of Nelson (1978) both worked on stations and habitations. The division made by Nelson (1978) – while practical in identifying two different area classifications – did differentiate between two fields. Nelson's history did dismiss ecological biogeography from the annals of modern historical biogeography (represented by Sclater's and de Candolle's regions), while Ratzel wanted to unite plant and animal geography in a general or an *Allgemeine Geographie*.

## Specialisation, Unity and the Proliferation of Terms

Plant and animal geography by the 1880s and biogeography in the 1970s were both multidisciplinary fields. While the fields of two different periods had radically different aims and methods, they did share one thing in common: a divergence of ideas and practises, while yearning for unification. The specialisation of biogeography occurred at least twice in the history of the field. With more researchers discussing plant and animal geography and as a consequence, a greater number of methods,

---

of Cox and Moore (2005), which is derived directly from (but not attributed to) Nelson (1978). It is quite clear, from the discussion above, that historical and ecological approaches "crystallised" in the work of nineteenth century plant geographers, rather than in the work early twentieth century animal ecologists.

approaches and theories, scientists were keen to encapsulate it under a single umbrella term – *biogeography*. But terms require definitions, and definitions call for syntheses. By the 1860s, plant geography was already divided into two distinct fields that used different classification systems to divide the worlds vegetation. A single umbrella term might encapsulate a large disparate group of researchers, but it did not necessarily define what they did. Neither Jordan nor Merriam, both of who coined the term "biogeography" in German and English respectively, defined their terms. However, those that were within "sub-disciplines" did. For example, Grisebach was careful to define *geobotanik* [geobotany] as "the vegetation of single countries" (Grisebach 1866, p. 373), which was grouped into topographical geobotany "the topographical character of vegetation based on the influence of surface processes", climatic geobotany "the influence of climate on the geographical arrangement of vegetation based on physiological conditions" and, geological geobotany [*sensu* the historical plant geography of Wulff]. Neither of these categories within Grisebach's classification includes the study of larger continental distributions (A. P. de Candolle's *habitations*).[13]

Grisebach was not the only plant geographer to professionalise their field. A.P. de Candolle (1820), A. de Candolle (1855), Schouw (1823) and Meyen (1836, 1846) attempted ways to distinguish what they did from other forms of geography. In reading these authors you can tell that the travelogues of the Forsters and Brown as well as the more "Gentleman naturalist" accounts by John Barton (1827), Edward Forbes (1846) and Hewett Cottrell Watson (1835, 1847–1859), undermined the sort of specialisation that people like Grisebach were seeing in geology and geography. For instance, in the German-speaking world, people were doing their doctorates in plant geography (e.g., Stromeyer 1800) and were being employed as specialists, that is, as botanists rather than as "naturalists". Perhaps this form of specialisation in the German speaking world had led to the export of German speaking plant and animal geographers abroad to places as far flung as Brazil, New South Wales and the Pacific. Home (1995) makes this point very clear "[nineteenth century] German universities were producing specialists, including of course specialists in various fields of science who were well trained to pursue scientific research of their own, in Britain, scientific and technical training remained a much more hap hazardous affair" (Home 1995, p. 17).

Specialisation also involves the unification of methods, goals and terms. While new terms were being defined to describe specialised areas, like *geobotanik*, others were attempting to unify these emerging fields under a single term *Allgemeine biogeographie* (Ratzel 1891). This proliferation of terms comes from those wanting to differentiate their work from others (topographical botany versus history plant geography) and, from those wanting to apply a synthesis to the study of plant

---

[13]Grisebach also challenges the Darwinian hypothesis with his own: "it is the opposite view that any particular organisation is the product of their living conditions, that individual plant species have emerged everywhere, in places where they were able to exist. The climate and soils create physical conditions that may exist in another areas, thereby leading to the existence of similar forms, with different levels of organisation, in different places" (Grisebach 1866, pp. 391–392).

and animal distributions (Darwinian biogeography versus geobotany). While Ratzel gave us biogeography to unify animal and plant geography, Haeckel proposed the terms "ecology" and "chorology" to distinguish between biogeographers, but few practitioners were using these terms and fewer were being unified.

The practise of classification had driven late eighteenth and nineteenth century plant geography. The classifications were highly divergent leading to plant geographers on both sides laying claim to a natural classification, one based on vegetation types and the other on taxa, like species genera and families. By 1859, Alexander von Humboldt, the person who started the concept of a classification of vegetation, had died. Ironically, it was the same year Darwin's *Origin of Species* appeared, which, together with the work of Wallace, was again to change both plant and animal geography, moving it away from geology and geography to the newly established field of biology.

New fields result in practitioners making new claims. In the late eighteenth century, plant and animal geography were lodged firmly within taxonomic practise, an appendix of a larger body of systematic work. Zimmermann and Stromeyer changed this by introducing the idea of geographical areas, classifying areas based on the species and genera that lived there. Humboldt moved plant geography into the realm of geography and latter nineteenth century botanists saw it as a biological field in which Darwin's ideas played an important part. The subdivision of plant and animal geography and the various fields that lay claim to it, did not have a great impact on its development.

Janet Browne is correct in her claim that "each discipline gave its practitioners a certain way of looking at things, a certain way of thinking about the evidence, and a distinctive way set of assumptions" (Browne 1983, p. 107). Specialisation arising in established fields, like geography and geology, often claim emerging scientific fields as "sub-disciplines". For example, evolutionary biologists of the twentieth century would group biogeography as a sub-discipline within evolutionary biology. Consider the statement by entomologist Phillip J. Darlington, the world's first self-professed professional biogeographer[14]: "evolution made modern zoogeography" (Darlington 1980, p. 28). Darlington's claim is possessive: biogeography is within the realm of evolutionary biology and *not* geography. To be fair to Darlington, he does consider biogeography in the post Darwin and Wallace era.

Charles Lyell made similar a claim in his third volume of the *Principles of Geology* (Lyell 1842), "there is another class go phenomena relating to the organic world, which have an equal claim on our attention, if we desire to obtain possession

---

[14]"If there is such a thing as a professional in biogeography, I am one. I am therefore in a position to know the complexities and difficulties of the subject. There are many" (Darlington 1965, p. 184), also "I am a conservative, sixty–year–old biologist and also a professional biogeographer (if there is such a thing)" (Darlington 1964, p. 1084). I am grateful to Gary Nelson for pointing out these two quotes. However, in a volume dedicated to the great entomologist, Ball (1985) states that Darlington calls himself a "professional biogeographer in his book *Evolution for Naturalists*" Ball (1985, p. 1). On closer inspection of *Evolution for Naturalists,* I found Darlington referring to himself as a "biologist" and not as a "biogeographer" (see Darlington 1980, p. x).

of all the preparatory knowledge respecting the existing course of nature, which may be available in the interpretation of geological monuments" (Lyell 1842, p. 1). Biology was not a profession in 1842 and geology could lay claim to the study of distributions through time, although few geologists practised plant and animal geography. Even so, it seems that geologists were happy to lay claim, as did geographers of the late nineteenth century, like Friedrich Ratzel and Georg Karl Cornelius Gerland, who grouped plant and animal geography within geography.[15]

Regardless where one places plant and animal geography, its aims and methods are driven by its practitioners, who were not necessarily allied with any field in particular. Ratzel may have claimed that it was "the duty of geography" to unify plant and animal geography, however few plant and animal geographers were practicing geographers.

The same is true for biogeography in the 1970s. Nelson divided biogeography into ecological and historical biogeography, the former ecologists and the latter practitioners of biological classification and phylogenetic systematics. Dividing a field into what the practitioners do creates a more realistic historical account of a field. In the case of ecological and historical biogeography, it created a dilemma: couldn't an ecologist work on regions? The problem was how the division was justified. Nelson used A.P. de Candolle's stations and habitations, or the difference between larger continuous distributions versus smaller local distributions. The problem was that fields such as island biogeography use both. One can equally divide biogeography into those who work on ahistorical or recent distributions and those who work on older historical distributions. Moreover, with the introduction of molecular data into biogeography you now had historical biogeographers working on very small areas. While the term historical biogeography is still used, it refers to non-ecological distributional research, that is, distributions through time.

Unity through specialisation has consistently failed to unify an ever increasingly divergent field. While plant and animal biogeography were driven by classification early on, it is practitioners, their skills and backgrounds, which is steering biogeography. For early plant and animal geographers it was how they used classification, the method of classifying the world rather than theories that drove plant and animal geography. Browne is right to say that "each discipline gave its practitioners a certain way of looking at things", but that is more relevant today than it ever was in the early nineteenth century.

---

[15]Gerland saw the role of "geophysics" (i.e., geological processes) as central to the study of geography. Unlike Carl Ritter and Julius Fröbel before him, Garland dismissed anthropocentric views. Humanity was not the product of the environment, but of society. Gerland's physical geography appears modern as it effectively links Earth processes with plant and animal geography, effective removing the human geographical elements. Leighly (1938) however did not see that link in Gerland's work "We can see know that plant and animal geography actually have no logical bond with the other components of Gerland's physical science of the earth. If they are excluded, the remainder is an accurate outline of geophysics ..." (Leighly 1938, p. 253). See Hartshorne (1939, pp. 286–288) for commentary.

# References

Allione, C. (1785). *Flora pedemontana sive enumeratio methodica stirpium indigenarum Pedemontii.* Turin: Joannes Michael Briolus.
Anonymous. (1856). VII. 1. Géographie Botanique raisonnée. Par M. Alphonse de Candolle. Paris: 1855. *The Edinburg Review, 104*, 490–518.
Aublet, F. (1775). *Histoire des plantes de la Guiane Françoise: rangées suivant la méthode sexuelle, avec plusieurs mémoires sur différens objects intéressans, relatifs à la culture & au commerce de la Guiane Françoise, & une notice des plantes de l'Isle-de-France.* Paris: P. F. Didot.
Ball, G. E. (1985). Introduction. In G. E. Balls (Ed.), *Taxonomy, phylogeny, and zoogeography of beetles and ants: A volume dedicated to the memory of Philip Jackson Darlington, Jr. 1904–1983* (pp. 1–3). Dordrecht: Dr W. Junk.
Barton, J. (1827). *A lecture on the geography of plants.* London: Harvey and Dalton.
Bradley, R. (1721). *Philosophical account of the works of Nature.* London: W. Mears.
Brady, R. H. (1972). Towards a common morphology for aesthetics and natural science: A study of Goethe's empiricism. Ph.D thesis, State University of New York, New York.
Browne, J. (1983). *The secular ark: Studies in the history of biogeography.* New Haven: Yale University Press.
Cox, B. C., & Moore, P. D. (2005). *Biogeography: An ecological and evolutionary approach* (7th ed.). Oxford: Blackwell.
Darlington, P. J. (1964). Drifting continents and late Paleozoic geography. *Proceedings of the National Academy of Sciences, 52*, 1084–1091.
Darlington, P. J. (1965). *Biogeography of the southern end of the world: Distribution and history of far-southern life and land, with an assessment of continental drift.* Cambridge, MA: Harvard University Press.
Darlington, P. J. (1980). *Evolution for naturalists: The simple principles and complex reality.* New York: Wiley.
de Candolle, A. P. (1805). Explication de la carte Botanique de la France. In J. B. P. A. de M. de Lamarck & A. P. de Candolle (Eds.), *Flore française, ou descriptions succinctes de toutes les plantes qui croissent naturellement en France, disposées selon une nouvelle méthode d'analyse, et précédées par un exposé des principes élémentaires de la botanique* (3rd edn, pp. v–xii). Paris: Desray.
de Candolle, A. P. (1813). *Théorie Élémentaire de la Botanique, ou Exposition des Principes de la Classification Naturelle et de l'Art de Décrire et d'Etudier les Végétaux.* Paris: Déterville.
de Candolle, A. P. (1818). Memoir upon the Geography of the Plants of France, considered more especially with Regard to their Height above the Level of the Sea. *Annals of Philosophy, 11*, 408–413.
de Candolle, A. P. (1820). *Essai élémentaire de géographie botanique. Dictionnaire des Sciences Naturelles* (Vol. 18, pp. 1–64). Paris: F. Levrault.
de Candolle, A. L. P. P. (1830). *Monographie des Campanulées.* Paris: Desray.
de Candolle, A. L. P. P. (1855). *Géographie botanique raisonnée.* Paris: Masson.
Drouin, J.-M. (1998). Augustin-Pyramus de Candolle. In P. Acot (Ed.), *The European origins of scientific ecology, Volume 1* (pp. 359–422). Amsterdam: Overseas Publishers Association.
Ebach, M. C. (2005). Anschauung and the archetype: The role of Goethe's delicate empiricism in comparative biology. *Janus Head, 8*, 254–270.
Ebach, M. C., & Goujet, D. F. (2006). The first biogeographical map. *Journal of Biogeography, 33*, 761–769.
Egerton, F. N. (1968). Studies of animal populations from Lamarck to Darwin. *Journal of the History of Biology, 1*, 225–259.
Forbes, E. (1846). On the connexion between the distribution of the existing fauna and flora of the British Isles and the geological changes which have affected their area, especially during the epoch of the Northern Drift. *Memoirs of the Geological Survey of Great Britain, 1*, 336–432.

Forster, J. R., & Forster, G. (1776). *Characteres generum plantarum, quas in itinere ad insulas maris Australis: collegerunt, descripserunt, delinearunt, annis 1772–1775*. London: B. White, T. Cadell, & P. Elms.
G.E.D. (1871). Geographisches Jahrbuch. *Nature, 4*, 44–45.
Gmelin, J. G. (1747). *Flora Sibirica sive Historia Plantarum Sibiriae*. Petropoli: Ex Typographia Academiae scientiarum.
Grisebach, A. H. R. (1866). Der gegenwärtige Standpunkt der Geographie der Pflanzen. *Geographisches Jahrbuch, 1*, 373–402.
Grisebach, A. H. R. (1872). *Die Vegetation der Erde nach ihrer klimatischen Anordnung*. Leipzig: Wilhelm Engelmann.
Hagen, J. (1986). Ecologists and taxonomists: Divergent traditions in twentieth-century plant geography. *Journal of the History of Biology, 19*, 197–214.
Hartshorne, R. (1939). The nature of geography: A critical survey of current thought in the light of the past. *Annals of the Association of American Geographers, 29*, 173–412.
Heer, O. (1835). *Die Vegetationsverhältnisse des südöstlichens Theils des Cantons Glarus; ein Versuch, die pflanzengeographischen Erscheinungen der Alpen aus climatischen und Bodenverhältnissen abzuleiten*. Zurich: Orell, Füssli.
Home, R. (1995). Science as a German export to nineteenth century Australia. *Working Papers in Australian Studies, 104*, 1–21.
Jackson, S. T. (2009). Introduction: Humboldt, ecology, and the cosmos. In A. von Humboldt & A. Bonpland (Ed.), *Essay on the geography of plants* (Jackson, S. T., Ed., pp. 1–52). Chicago: University of Chicago Press.
Jax, K. (2011). Stabilizing a concept. In A. Schwarz & K. Jax (Eds.), *Ecology revisited: Reflecting in concepts, advancing science*. New York: Springer.
Kinch, M. P. (1974). An assessment of rival British theories of biogeography, 1800–1859. M.Sc. thesis, Oregon State University, Corvallis.
Leighly, J. (1938). Methodologic controversy in nineteenth century German geography. *Annals of the Association of American Geographers, 28*, 238–258.
Lorenz, J. R. (1863). *Physicalische Verhältnisse und Vertheilung der Organismen im Quarnerischen Golfe*. Wien: Die kaiserlich-königliche Hof- und Staatsdruckerei.
Lyell, C. (1833). *Principles of geology: Being an attempt to explain the former changes of the earth's surface, by reference to causes now in operation* (Vol. 2). London: J. Murray.
Lyell, C. (1842). *Principles of geology or the modern changes of the earth and its inhabitants, considered as illustrative of geology by Charles Lyell* (Vol. 3). Boston: Hilliard, Gray & Co.
Meyen, F. J. F. (1836). *Grundriss der Pflanzengeographie mit ausführlichen Untersuchungen über das Vaterland, den Anbau und den Nutzen der vorzüglichsten Culturpflanzen, welche den Wohlstand der Völker begründen*. Berlin: Haude und Spenersche Buchhandlung.
Meyen, F. J. F. (1846). *Outlines of the geography of plants*. London: Ray Society.
Moss, C. E. (1910). The fundamental unites of vegetation. *New Phytologist, 9*, 18–53.
Nelson, G. (1978). From Candolle to Croizat: Comments on the history of biogeography. *Journal of the History of Biology, 11*, 269–305.
Nicolson, M. (1996). Humboldtian plant geography after Humbodlt: The link to ecology. *British Journal for History of Science, 29*, 289–310.
Nordenskiöld, E. (1936). *The history of biology: A survey*. Translated from the Swedish by Leonard Bucknall Eyre. New York: Tudor.
Nyhart, L. K. (2009). *Modern nature: The rise of the biological perspective in Germany*. Chicago: University of Chicago Press.
Oeder, G. C. (1761). *Flora Danica*. Kopenhagen: C. et A. Philibert.
Pallas, P. S. (1784–1788). *Flora Rossica*. Petropoli: J. J. Weitbrecht.
Ratzel, F. (1891). *Anthropogeographie* (Vol. 2). Stuttgart: Engelhorn.
Schmarda, K. L. (1853). *Die geographische Verbreitung der thiere*. Vienna: Carl Gerold and Son.
Schmarda, K. L. (1866). Die Theirgeographie und ihre Aufgabe. *Geographisches Jahrbuch, 1*, 402–427.
Schouw, J. F. (1823). *Grundzüge einer allgemeinen Pflanzengeographie*. Berlin: Reimer.

Sendtner, O. (1854). *Die Vegetationsverhältnisse Südbayerns nach den Grundsätzen der Pflanzengeographie und mit Bezugnahme auf Landeskultur.* München: Geschenk des Verfassers.
Shimwell, D. W. (1971). *The description and classification of vegetation.* London: Sidgwick & Jackson.
Stevens, P. F. (1994). *The development of biological systematics: Antoine-Laurent de Jussieu, nature, and the natural system.* New York: Columbia University Press.
Stott, P. (1984). History of biogeography. In J. A. Taylor (Ed.), *Biogeography: Recent advances and future directions* (pp. 1–24). Totowa: Barnes & Noble Books.
Stromeyer, F. (1800). *Commentatio inauguralis sistens historiae vegetabilium geographicae specimen.* Göttingen: Heinrich Dieterich.
von Goethe, J. W. (1790). Versuch die Metamorphose der Pflanzen zu erklären. Gotha: Ettingersche Buchhandlung.
von Goethe, J. W. (1813). Höhen der alten und neuen Welt bildlich verglichen. *Allgemeinen Geogrphischen Ephemeriden, 41*, 3–8.
von Humboldt, A. (1793). *Florae Fribergensis specimen.* Berlin: H. A. Rottmann.
von Humboldt, A., & Bonpland, A. (2009). *Essay on the geography of plants* (S. T. Jackson, Ed.). Chicago: University of Chicago Press.
Von Marilaun, A. K. (1863). *Das Pflanzenleben der Donaulander.* Innsbruck: Wagner.
Watson, H. C. (1835). *Remarks on the geographical distribution of British plants.* London: Longman, Rees, Orme, Brown, Green, and Longman.
Watson, H. C. (1847–1859). *Cybele Britannica; Or British plants and their geographical relations* (4 vols.). London: Longman & Co.
Whittaker, R. H. (1962). Classification of natural communities. *Botanical Review, 28*, 1–239.
Wilkins, J. S. (2009). *Species: A history.* Berkeley: University of California Press.
Wilkins, J. S., & Ebach, M. C. (2014). *The nature of classification: Relationships and kinds in the natural sciences.* Basingstoke: Palgrave Macmillan.
Willdenow, C. L. (1805). *Grundriss der Kräuterkunde.* Berlin: Haude and Spener.
Willdenow, C. L. (1811). *The principles of botany and of vegetable physiology.* London: William Blackwood.
Williams, D., & Ebach, M. C. (2008). *Foundations of systematics and biogeography.* New York: Springer.

# Chapter 5
# Plant and Animal Geography in Practise: Maps, Regions and Regionalisation

> *We have now endeavoured to demonstrate the insufficiency of all theories on the causes of animal dispersion, and yet experience teaches us, that certain divisions of the earth are characterised by peculiar animals. We are now to enquire, what are these divisions? how they are to be defined? and what are their peculiarities? (Swainson 1835, pp. 9–10).*

No matter how taxonomic and vegetational classifications were justified, whether by distributional laws or by latitude, they become trivial once we look at *how* plant and animal geographers divided up the natural world as areas. Rather than mapping or analysing distributions, distributional laws help explain why they are there. In other words, regionalisation is about distribution and not about distributional laws and theories.

Many historians avoid the history of regionalisation. It is a messy subject that relies heavily on existing classifications and known distributions (or distributional models). While historians like to investigate the distributional laws and theories – dispersals from mountaintops after biblical floods or by elaborate rafting events across vast oceans (Bowen 1983) – they did not take up a great deal of the practitioners' time. Theories and laws are ideas; some ideas that may appear in an introduction to Hooker's *Flora Antarctica* (Hooker 1844–1859), or in Linnaeus' *Flora Lapponica* (Linnaeus 1737), but the crux of each of these geographical studies was distribution – where do these taxa occur? More important, do they fall within pre – defined geographical regions?

In this chapter I wish to show that regionalisation is dependent on distributions, classification and methodology rather than on distributional laws. Moreover, I wish to show that similar classifications can produce different regionalisations and how practitioners of plant and animal geography have had a hand in shaping regionalisation in the latter half of the nineteenth century.

## Mapping Distribution (1777–1800)

It is rare that distributional laws, let alone theory, shape how we practise our field. Take taxonomy for example; no taxonomist has a species concept in mind when they describe their taxa (see Wilkins and Ebach 2014). The taxonomist is focused on describing a specimen and comparing it against many other specimens. No theoretical process is assumed, just comparisons between the parts of individual specimens. The same is true when early plant and animal geographers drew maps and proposed animal and plant regions. The regions were not based on distributional laws or mechanisms. Rather they were based on distributions of taxa and physical geography (e.g., mountain chains, the position of oceans, lakes and seas and isotherms). The first example of a distributional map is that of Zimmermann's 1777 *Tabula mundi*, which consists of quadruped distributions drawn onto a world map with a Mercator projection. The map contains both place names as well as the locations of mountain chains, seas and oceans. According to Schmithüsen (1985, p. 66) and Feuerstein-Herz (2004, p. 233), the goal of the map was to be able to view the whole animal world in one go.[1] In his 1783 "Short explanation of the zoological world map", namely the 1777 *Tabula mundi*, Zimmermann states:

> This attempt at a zoological world map not only shows at a glance how many quadrupeds are known to date, but it determines their living place [Wohnplatz]. It can therefore serve zoologists as a resource to find an animal in their own country (Zimmermann 1783, p. 3. my translation).[2]

Zimmermann's emphasis on "Wohnplatz"[3] is important. While Schmithüsen and Feuerstein-Herz may have missed its relevance as it places greater weight for the argument that distributional maps are way to identify the "Wohnplatz" of known and named animals.

The "Wohnplatz" can be loosely translated as the "living place", rather than a "location" (i.e., Ort, Platz etc.), of an organism. This is not to suggest Zimmermann was talking about endemism, a concept that came later in the nineteenth century. Rather, Zimmermann noticed that certain organisms are found in places *to which*

---

[1]Hofsten has a seemingly presentist take of Zimmermann's map: "I can only discuss general principles and conclusions here; if one disregards all outdated explanations and emphasises the important viewpoints of today, his map may be summarised as follows: distribution is controlled by climate" (Hofsten 1916, p. 253, my translation). Hofsten conclusions may appeal to a presentist perspective of eighteenth century animal geography, however, the interpretations of Schmithüsen (1985) and Feuerstein-Herz (2004) are, in my view, historically accurate.

[2]The original reads: "Dieser Versuch einer zoologischen Weltcharte zeigt nicht nur auf einen Blick, wie viel Quadrupeden bis jetzt uns bekannt sind, sondern sie bestimmt jeder Art ihren Wohnplaz [sic]. Sie kann daher den Zoologen in so weit als ein Hülfsmittel dienen, ein Thier sogleich in seinem Vaterland aufzufinden" (Zimmermann 1783, p. 3).

[3]Throughout Zimmermann's three volume *Geographische Geschichte des Menschen, und der allgemein Verbreiteten Vierfüßigen Thiere* (Zimmermann 1778–1783), the misspelt "Wohnplaz" appears in the first two volumes, while "Wohnplatz" only appears in volume three. The latter usage is correct in modern German and is used herein.

*they belong*. For example, Zimmermann cites Buffon when discussing the "living place" of tigers (Zimmermann 1780, p. 260). Buffon, however, does not use an equivalent term to "Wohnplatz", rather opting for "confined to" and "regions or lands [contrées]" (Buffon 1761, v. 9, p. 131). The idea of a Wohnplatz is significant as it means organisms are not necessarily *confined to*, but rather that they *live in* or *belong to* a place. While this would have a lasting impact on animal regionalisation, Zimmermann did not translate the idea of "Wohnplatz" to his maps, rather opting for boundaries [Grenzen], countries [Lander *sensu* Buffon's contrées] and, continents [Weltteile]:

> You can not accurately sketch out the living place [Wohnplatz] of an animal species, so that you can show the sum of the square miles in which they only live. This can be done with far less uncertainty if we use the degrees of latitude and longitude[4] (Zimmermann 1780, p. 3, my translation).

Zimmermann, it seems, went for the Linnaean option of latitude and longitudes, that is, for an artificial classification to help identify areas, rather for an approach that would describe natural (or even climatic) areas.[5] However, "an exact size specifying the extent of the inhabited areas was, as Zimmermann realised, impossible to achieve" (Feuerstein-Herz 2004, p. 199). It seems unreasonable to dismiss Zimmermann.[6] Not much was known of organismal distribution and many well species had incomplete distributions. Moreover, Zimmermann, like many zoologists at the time, were working from Buffon's *Histoire*, which itself contained many inaccuracies.[7]

The only other contemporaneous distribution map to Zimmermann's was the *Geographical Map of Nature. Or the Natural Distribution of Minerals Plants etc.*

---

[4]The original reads "So genau kann man aber keiner Thierart die Grenze ihres Wohnplazes vorzeichnen, daß man die Summe von Quadratmeilen, binnen welcher sie nur leben, angeben wollte. Nach den Graden der geographischen Länge und Breite läßt sich dies mit geringerer Unbestimmtheit thun" (Zimmermann 1780, p. 3).

[5]A year after Zimmerman's *Tabula mundi*, Danish entomologist Johan Christian Fabricius published his *Philosophica Entomologica* (Fabricius 1778). In it, Fabricius divides the world into eight climatic regions "from which the Stations of insects are judged" (Fabricius 1778, p. 154). The eight regions – 1. Indian, 2. Egyptian, 3. Austral, 4. Mediterranean, 5. Boreal, 6. Oriental, 7. Occidental and, 8. Alpine – differ from Zimmerman's regions as they are climatic, however, each region is described by country (i.e., "Boreal, Europe between Lapland and Paris"). Fabricius unfortunately did not produce a map. French zoologist Pierre André Latreille, however, considered Fabricius' regions arbitrary: "Ce simple exposé suffit pour nous convaincre qu'il y a dans ces divisons beaucoup d'arbitraire" (Latreille 1815, pp. 40–41), described by Belgian entomologist Jean Théodore Lacordaire as "une autre division beaucoup meilleure" (Lacordaire 1834, p. 600).

[6]Zimmermann did include the most northerly and southerly limits of camels, reindeer, moose and elephants. Had he dared encapsulate these, he would have possibly produced the first ever area polygon delimiting the distribution of genera.

[7]Buffon suggested that the tiger lives in the same lands [contrées] as the elephant and rhinoceros (Buffon 1761, p. 131). To the modern biogeographer this may conjure up a scene from Monty Python's *Meaning of Life*: "A tiger? In Africa?"

*observed in Vivarais*[8] published in *Géographie de la Nature* by Jean-Louis Giraud Soulavie (1780). This was similar in composition to that of Zimmermann. The map of the Vivarais region in south-east France was the "first map of the [Vivarais] Province that has been brought to light and is modelled on my relief map and that of the Academy" (Giraud Soulavie 1780, p. 16, my translation). The map contains various pieces of information: the contour or relief of the region, which encompasses the drainage basins of the Loire and Rhône, as well as the volcanic soils (highlighted in red), and limestone soils (highlighted in blue). The map contains the "Kingdom [Règne] of Alpine plants" (marked as a symbol – .·.), the locations of minerals (e.g., arsenic, iron, lead, quartz), the limits of vines and volcanic craters. As a distribution map it seems rather general and less specialised to that of Zimmermann. Giraud Soulavie does however include a list of nine "facts" and "observations" regarding plant geography in his "Géographie de la Nature": "Let us establish the facts and observations that explain what we mean by plant geography" (Giraud Soulavie 1780, p. 11). Regardless, Giraud Soulavie's map does not include any of these facts or observations, again showing a disconnection between theory and practise. For example, Giraud Soulavie's mentions a method for studying plant geography based on his nine points. Using meteorological instruments, one can measure the decrease in the atmospheric warmth as you climb from the base of the mountains to the top, from which you can observe plants, which have "chosen a similar climate to suit their constitution" (Giraud Soulavie 1780, p. 13). In other words, altitude controls plants distribution. While the method is perfectly valid, it is not reflected in his *Geographical Map of Nature* [*Géographie de la Nature*]. However, Giraud Soulavie and his *Géographie de la Nature* did attract the attention of one plant geographer in particular:

> [*Géographie de la Nature*] contains some and profound and ingenious views as to the forms, relations, and habitudes of vegetables (Humboldt 1816, p. 446).

Humboldt is referring to the seven tome 1780–1784 work, *Histoire naturelle de la France méridionale*, in which Giraud Soulavie (1783a, b) presents a cross section of Mount Mezin in the Loire Valley, the same area featured in his 1780 work (Fig. 5.1). The map titled *Vertical cross-section of the Vivaroises Mountains; respective limits of plants*, was no different to that of Goethe's "Heights of the Old and New World, figuratively compared", in which altitudinal lines are used to delimit types of climates (e.g., "Alpine Climate", "Climate of Chestnuts" "Climate of Vines"; Giraud Soulavie 1783a, b). The practise of using a cross section to indicate climatic zones (and plant geography) was also adopted by the South American cartographer, Francisco José de Caldas, also used cross sections, but to greater effect. In his unpublished work *Memoir on the distribution of plants that are cultivated near the equator*, de Caldas produced a series of intricate, and accurate cross sections

---

[8]The original reads: "Carte Géographique de la Nature. Ou disposition Naturelle des Minéraux Végétaux &c observée en Vivarais". Vivarais is a traditional region in the south-east of France.

(or "phytogeographical profiles") of Nueva Granada (New Kingdom of Granada), which were colourised according to plant region:

> I have been able to make within the Viceroyalty of Santafé, my first concern has been to observe the altitude, the quantity and the limits which mark the cultivation of useful plants and on which we depend for our subsidence. Since 1796, when I began to reflect upon these matters, until today (April 1803), I have compiled a considerable number of observations and facts [...] The accompanying plan represents (sideview) [cross section] all of the terrain covered by my observations: it starts from 4°36′ northern latitude to 0° southern latitude; that is to say, from Santafé [Bogota] to Quito. (de Caldas translated in Appel 1994, p. 139).

The phytogeographical profiles are striking. Like the cross section of Giraud Soulavie, Caldas' maps are based on far more data, mostly barometric readings with altitudes and latitudes (Appel 1994, p. 57). The profiles however also contain plants regions based on economical crops, such as "Región de la *Theobroma cacao*", the Region of the cacao plant and, "Región del género *Musa*", the region of the Banana genus. Caldas and Giraud Soulavie's maps serve the same purpose, the delineated where important crops are found and the geographical conditions (e.g., climate, altitude) that support them (see also Bourguet 2002, Güttler 2015[9]). The cross sectioning methods of both Caldas and Giraud Soulavie were seemingly adopted by Humboldt, although Appel (1994) argues differently:

> Could it be that Humboldt regarded Caldas as a rival?[10] I believe that for Humboldt the concept of phytogeography was especially important. He found in Nueva Granada two men, Mutis and Caldas, capable of developing the idea, and one, Caldas, already with the raw data available to do it (Appel 1994, p. 59).

This does not follow from the viewpoint of a biogeographer; Caldas *was* already doing phytogeography. Humboldt just refined it to define natural vegetation types, rather than commercial crops.

Humboldt however changed the way eighteenth century plant geography was done by focusing on natural vegetation types, rather than purely economically important crops. Moreover, Humboldt's adopted methodology had few takers. Goethe, upon noticing that Humboldt's *Tableau* was missing from the 1807

---

[9]Güttler claims, "for early plant geographers, the distribution borders of crops gave an indication of how wild plants might also be distributed in space" (Güttler 2015, p. 30). While this is certainly true for Giraud Soulavie and Caldas, who mapped cultivated plants, however it is not true for Humboldt or A.P. de Candolle.

[10]Appel thinks not. There has been some controversy whether Humboldt stole Caldas' ideas (Serje 2004). Jackson (2009), however, refutes this claim based on Humboldt's earlier work and his reading of Ramond de Carbonnières (1798) (Jackson 2009, p. 13, footnote 20 & p. 246). I am inclined to disagree. Humboldt's earlier work was merely a preamble to a geography of plants and did not include a methodology, certainly not a phytogeographical profile. Humboldt's *Tableau* and Caldas' phytogeographical profile are too similar in their construction (and aim) to be of coincidence. However, one could argue that idea of a phytogeographical profile had universal appeal, as Giraud Soulavie's cross section shows.

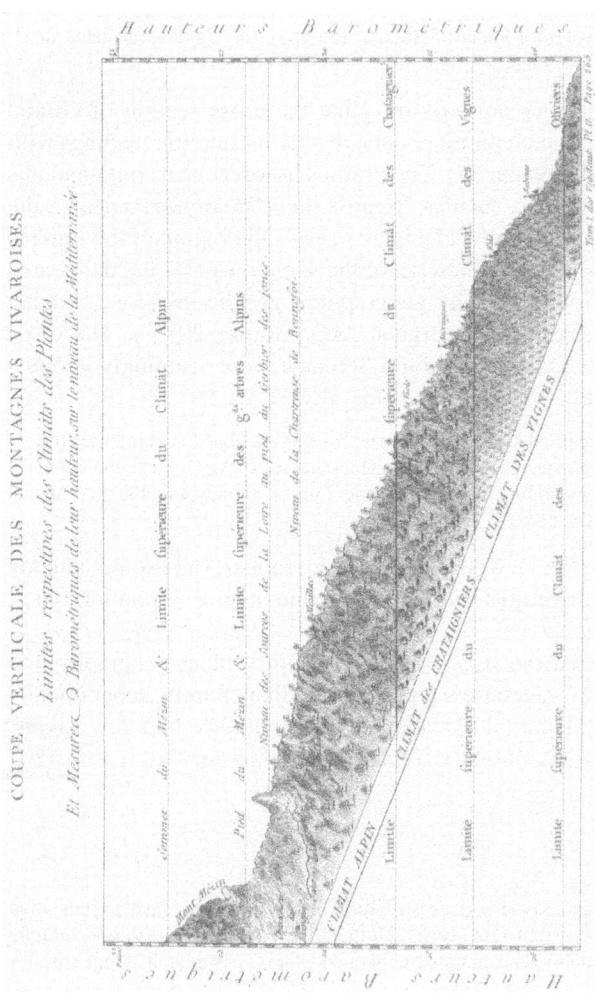

**Fig. 5.1** *Vertical cross-section of the Vivaroises Mountains; respective limits of plants* by Giraud Soulavie (1783a, b, Vol. 1, Part II, "The Vegetables", p. 265 original italics). The original reads COUPE VERTICALE DES MONTAGNES VIVAROISES. *Limites respectives des Climâts des plantes. Et mesures Baromètriques de leur hauteur, sur le niveau de la Méditerranée*. Note how the plant areas are divided by altitude. From the top: *Summit of Mount Mézin and the upper limits of the Alpine climate* [*Sommet du Mézin & Limites supérieure du Climât Alpin*]; *The foot of Mount Mézin and the upper limit of Alpine trees* [*Pied du Mézin & Limite supérieure des g$^{ds}$ arbes Alpins*]; *Level of the source of the Loire at the foot of the Gerbier sedges* [*Niveau des Sources de la Loire au pied du Gerbier des joncs*]; *The Chartreuse de Bonnefoy level*; *Upper climatic boundary of chestnuts (Châtaigniers)* [*Niveau de la Chartreuse de Bonnefoi; Limite supérieure du Climât des Chataignier$^s$*]; *Upper climatic boundary of olives* [*Limite supérieure du Climât des Vignes*; *Limite supérieure du Climât des Oliviers*]. The axis on the cross-section is barometric height (*Hauteurs Baromètriques*). Altitude is between 26 and 22 "Pouces" or inches (Source: UNSW Australia Library)

German edition, drew his own based on Humboldt's method. The cross sectioning method (a.k.a. phytogeographical profiling) was most likely too complicated for many botanists as it required being present at the area of study and taking many measurements. Few had the resources or the time to do this, other than Humboldt. Augustin de Candolle, used a similar methodology to delimit natural areas in France based on written accounts as well as his own travels. This resulted in a larger map that covered a much wider area (i.e., France, Switzerland, and surrounding countries and dukedoms). The use of measurement to draw natural regions became increasing difficult as the area increased in size. While it was possible to map the vegetation regions between Santafé and Quito, it required a vast amount of resources that practicing plant geographers did not have. Moreover, in areas such as Australia, in which much of the landscape and flora were unknown, it was impossible to guess at what actually might be there other than a general description of the climate or coastline. Natural regions too were large, much larger than Caldas' Nueva Granada or Giraud Soulavie's Loire Valley. Humboldt most likely realised this, perhaps explaining why he only used his method once. But the phytogeographical profiles were not the only method. The practise of classifying vegetation types based a single taxon (e.g., cacao species or banana genus) really started off vegetation geography. Humboldt never states where the concept of a vegetation came from – perhaps it was from Caldas or Giraud Soulavie or both. But a vegetation classification is possibly the most important turning point in plant geography, as it defines nineteenth century phytogeography and early twentieth century ecology.

Agriculture and mapping economically important crops is not a surprising origin for nineteenth century plant geography and ecology. As a way to measure the environmental conditions to classify vegetation regions in which commercially viable crops can grow, seems a sensible idea. However, natural regions are a completely different concept. It required, among other things, a mechanisms. Humans introduce crops to new areas, but naturally occurring floras and faunas are not. Much of the historical literature is dedicated to exploring natural mechanisms for distribution, like dispersal and centres of origin (see Bowen 1983). In this chapter we are interested in how the concept of a natural area, one that can be mapped and used by practitioners, was developed in both plant and animal geography.

## Mapping Natural Areas (1805–1858)

The concept of a natural area was based on plant distributions that were delimited by nature (i.e., soils, climate, altitude etc.). While most plant geographers in 1800 understood this, would they rather outline the barriers to distributions as Humboldt did with his *Tableau*? This geographical approach of measuring and delimiting the isotherms, soil types and rainfall that control vegetation (i.e., plant forms) and their distributions, more taxonomically inclined plant geographers sought to find the regions of distributions. The first to do this was Augustin de Candolle, in his "Botanical Map of France" (de Candolle 1805). Treating areas as analogs to taxa means that you can identify, name and describe natural regions of distributions. De Candolle attempted to map these regions on his Botanical Map but was careful not to draw any exact boundaries:

> On this map, France is divided into five regions distinguished by different colours; one should note that these areas are not etched into nature exactly as they appear here. It would be difficult to represent their exact delimitation therefore these regions have to be considered only as very generalised indicators (de Candolle in Lamarck and de Candolle 1805, p. vi, translated in Ebach and Goujet 2006, p. 766).

By 1805 plant geography had two very different approaches. A geographical approach, pioneered by Humboldt, which involved measuring geographical phenomena, such as altitude, temperature and rainfall, and mapping these in relation to vegetative forms. The other approach, introduced by de Candolle, sought to identify, name and describe regions of distribution, analogous to taxa in taxonomy. Here we have an interesting case for a type of specialisation in plant geography, namely taxonomy (a term de Candolle later coined in 1813). A.P. de Candolle was an early taxonomist and used a taxonomic method (i.e., identifying, naming and describing natural phenomena), while Humboldt had a geographical background and implemented a geographical method (i.e., using instruments to measure geographical phenomena). The result of these two approaches can be seen in how the maps are drawn. Humboldt gives us a profile, with elevations and climatic zones on to which the position of individual species, genera, and regions of genera are mapped as names, and not as polygons (Table 5.1). Given Humboldt's labour and data intensive approach, it provides little in terms of defining regions on maps. Regardless, both approaches aim to identify natural areas, only de Candolle's approach however, actually maps regions as definable and abutting polygons.

Wahlenberg (1812) in his *Flora Lapponica* also included a geographical map (*Mappa botanico-geographica, tabula temperaturae et tabulis botanicis*), which featured his plant regions as colourised polygons (Wahlenberg 1812, p. LXVI, tab. 1; see also Mennema 1985, p. 117, Fig. 5.2). A year later in his *De vegetatione et climate in helvetia septentrionali inter flumina rhenum et arolam* Wahlenberg (1813) included a phytogeographical profile, similar to that of Humboldt's *Tableau* (Wahlenberg 1813, tab. 1).[11] Wahlenberg was foremost a Linnaean taxonomist who employed both approaches. Schouw, however, following Humboldt's classification of vegetation types, did use polygons to describe plant distribution (Mennema 1985, p. 117). Berghaus (1845), Forbes (1854) and Merriam (1899) all used Humboldt's lateral or transverse profiles. The use of polygons or isothermal lines to indicate distributions was far more popular. Humboldt's method, namely his *Tableau*, had

---

[11]It is also worth noting that A.P. de Candolle, Wahlenberg and Schouw had a critic, botanist Hewett C. Watson, famous for his *Cybele Britannica* (Watson 1847–1859). Watson did not believe in natural plant areas "nature does not admit of precise boundary lines" (Watson 1836, p. 21; see also Güttler 2015) and preferred plant geographers to map single taxon distributions. Moreover, Watson was critical of the maps produced by A.P. de Candolle, Wahlenberg and Schouw, which he claimed, "do not exhibit the name of a single plant" (Watson 1836, p. 18), even though his own map *The distribution of British Plants* (Watson 1847) did not contain any plant names.

Mapping Natural Areas (1805–1858)

**Table 5.1** Selection of regionalisations used by plant and animal geographer between 1777 and 1858

| Publication | Geographical region | Climatic region | Natural region |
|---|---|---|---|
| Zimmermann (1777)* | Continents & countries | – | – |
| Fabricius (1778) | – | Climatic regions | – |
| Forster (1778) | Continents & countries | – | – |
| Willdenow (1792) | Continents & countries | – | – |
| Giraud Soulavie (1783a, b)* | – | – | Regions of vegetation |
| Stromeyer (1800) | Old & New worlds | – | – |
| de Caldas 1802* | – | – | Regions of vegetation |
| de Candolle (1805)* | – | – | Plant regions |
| Humboldt & Bonpland (1807)* | – | Climate/vegetation | – |
| Wahlenberg (1812, 1813)* | – | – | Regions of vegetation |
| Illiger (1815) | Continents & countries | – | – |
| de Candolle (1817) | – | – | Plant regions |
| Humboldt (1817) | – | Isothermal lines | – |
| de Candolle (1820) | – | – | Plant regions |
| Schouw (1823)* | Continents & countries | – | Regions of vegetation |
| Prichard (1826) | – | – | Animal regions |
| Minding (1829) | Continents & countries | – | – |
| Watson (1835, 1847—1859) | Continents & countries | – | Regions of vegetation |
| Meyen (1836) | Continents & countries | – | Regions of vegetation |
| Berghaus (1845)* | Continents | Isothermal lines | Provinces |
| Schmarda (1853)* | – | – | Animal Regions |
| Agassiz (1854)* | – | – | Realms |
| Forbes (1854)* | – | Homoiozoic Belts | – |

An asterisk (*) denotes accompanying map. For a more complete list of regionalisations used in animal and plant geography see Merriam (1892, p. 7–20) and Arldt (1906, pp. 220–222, Tables I–III)

no enduring legacy in the nineteenth century.[12] Giraud Soulavie pioneered the use of isothermal lines, however, the use of naturally occurring vegetation types was Humboldt's one lasting legacy, which is still with us today, although heavily modified.[13] While plant geography may be divided into taxonomic distribution maps and vegetation distribution maps, what of animal distribution maps? Swainson summed this up nicely:

---

[12] Berghaus however uses Humboldt's phytogeographical profile to depict Tenerife, Andes and the Himalayas in his *Umrisse Der Pflanzengeographie* in the fifth volume of his *Physikalischer Atlas* (Berghaus 1838).

[13] Robinson and Wallis (1967) however, claim that Humboldt's isothermal map of 1817 was "an event of first rank in the history of thematic cartography" (Robinson and Wallis 1967, p. 122).

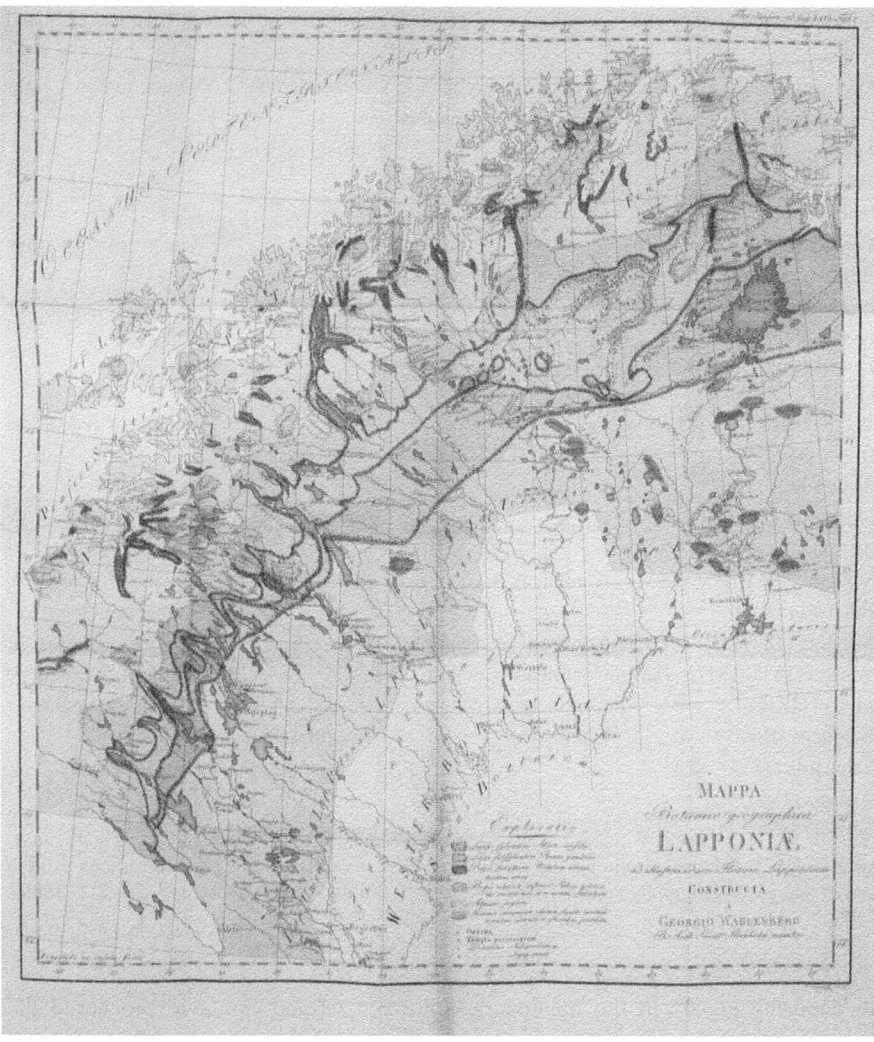

**Fig. 5.2** Wahlenberg's *Mappa botanico-geographica, tabula temperaturae et tabulis botanicis* (1812)

On looking at a map of the world we inhabit, we find that its surface is divided between land and water, continents and oceans; each, for the most part, thrown together into vast masses, placed under different temperatures, peopled by different races of men, and inhabited by peculiar sorts of animals. Two questions then occur to the mind. What are the causes that have produced this dissimilarity of creatures? and, secondly, *is there method in all this amazing diversity*? Each of these questions is highly interesting, and demands a separate consideration (Swainson 1835, pp. 1–2, my italics).

For the most part this chapter deals with the idea of a *method*, rather than an explanation for the "dissimilarity of creatures". The method, regionalisation, is what is used to divide the work into arbitrary or natural units. It is something we can use, regardless whether we believe that life started on Mount Ararat by divine deluge or naturally through transmutation. The explanation is secondary to the classification (see Wilkins and Ebach 2014) and has no impact on we practise plant and animal geography.

The method Swainson refers to is a classification. After all, his 1835 book *The Geography and Classification of Animals*, Swainson spends a great number of pages discussing artificial and natural methods and also marks a milestone in the development of a natural classification (Williams and Ebach 2008). No respecting taxonomist would knowingly use an artificial method in their classification, which would suggest the same for animal geography. In any case, Swainson remains silent about what these methods are or how they are implemented. Swainson however does review the insect regionalisations of Fabricius and Pierre André Latreille. Fabricius, he notes,

> ... by not attempting to demonstrate the correctness of any one of his divisions, seems to have subsequently abandoned them altogether, since no one, it may be fairly presumed, was more qualified than himself to discover the artificial nature of his theory (Swainson 1835, pp. 10–11).

However, Swainson, citing English entomologists William Kirby and William Spence (1826), dismisses the provinces of both Fabricius and Latreille, "built upon climate and temperature" and fixed "by degrees of longitude and latitude" as very similar approaches that are both artificial. What of natural regions? Swainson praises fellow Englishman James Cowles Prichard:

> Dr. Prichard is the first who attempted a more natural theory of animal distribution. This intelligent writer has looked more to the configuration of the earthy and to the natural connection or separation of its parts by intervening islands or oceans, than to absolute limits of longitude or latitude. Accordingly, from this very circumstance, his zoological divisions are formed with much greater attention to nature than any of his predecessors (Swainson 1835, pp. 12–13).

I point out Swainson's claim as it goes against much of the thinking within plant geography during the mid-nineteenth century. By 1835, the method of applying longitude or latitude (and altitude) had become common practise among the Humboldtian plant geographers, such as Meyen (1836) as well as marine geographers like Forbes (1846). Again, this may stem from the fact that Swainson is tackling animal regions from a taxonomic angle. For example, Swainson points out that Prichard's "arctic regions of the New and Old world" may suffer from analogies when searching for affinities. That is the "arctic regions of America, Europe, and Asia, indisputably possess the same genera, and in very many instances the same species; and if it should, subsequently appear that these regions are sufficiently important in themselves to constitute a zoological province, then it is a perfectly natural one; for not only are the same groups, but even the same species, in several instances, common to both". Swainson continues, "But can this be said of the second of these provinces, made to include the temperate regions of three continents?" (Swainson 1835, p. 13; also see Williams and Ebach 2008, p. 246).

Where the Humboldtians see natural areas of vegetation, the taxonomists see analogies. For Swainson, the temperate regions of America, Europe, and Asia were artificial or analogous. Compare this to Meyen's "Warmer Temperate Zone":

> This zone embraces the space between 34° and 45° of latitude; in Europe, including the flora of the south of Europe, as far as the Pyrenees, the mountains in the south of France, and those in the north of Greece. Asia Minor, the tract between the Black Sea and the Caspian, the northern part of China, and Japan lie in this zone (Meyen 1836, translated in Meyen 1846, p. 191).

The glaring differences between animal and plant geography were obvious by the beginning of the nineteenth century. Taxonomists sought a natural classification based on taxa, while the more geographically inclined sought climatic zones based on latitude and longitude.

What is surprising at first is that English zoologists like Prichard and Swainson and French botanists such as A.P. de Candolle introduced natural regions based on natural taxonomic classifications. The driver was clearly natural classification, something that de Candolle and Swainson both championed.[14] Given the natural animal regions proposed by Prichard (1826) and Swainson (1835), the first nineteenth century animal distribution map didn't appear until Berghaus (1845).

German cartographer Heinrich Berghaus was possibly the first to publish highly detailed distributions of animals.[15] The distributions drawn as lines as in Zimmermann's 1777 *Tabula mundi* depict the limits of mammal distribution. Several of his maps also contained cross-sections that are reminiscent of Humboldt, depicting the altitudes of mammals. Interestingly, Berghaus makes the distinction between geographical "Verbreitung" (extent or distribution) and "Verteilung" (division) of mammal distributions. While the German terms "Verbreitung" and "Verteilung" are clearly analogous, the respective maps of Berghaus are not. For example, the 1845 map titled *Vertheilung der Nagethiere und Wiederkäuer* (Division of the Rodents and Ruminants), contains 14 zoological provinces that are missing on the other distribution maps, *Geographische Verbreitung* (Geographical distribution), *Verbreitung der vorzüglicheren Saugethiere der Alten Welt* (Distribution of a selection of mammals of the Old World) and *Verbreitung der vorzuglicheren Saugethiere der Neuen Welt* (Distribution of a selection of mammals of the New World) all

---

[14]Latreille, however, also attempted a natural classification: "[b]y 'method' he and his contemporary naturalists mean the apparent tabulated results of a classification, whatever the approach leading to such a statement. For Latreille this method has to be natural" (Dupuis 1974, p. 7). I am not sure whether this constitutes as a natural method. The "philosophical" musings of de Candolle (1813) and Swainson (1835) did see a dramatic shift toward natural classification, one that had revolutionised systematics by the end of the nineteenth century (see Williams and Ebach 2008).

[15]Karl Ritter's *Sechs Karten Von Europa* (Ritter 1806), contains a phytogeographical and map titled *Tafel der wildwachsenden Bäume und Sträuche in Europa* and *Tafel über die Verbreitung der gezähmten und wilden Saugetiere in Europa* respectively. Unlike the maps of Berghaus, Ritter's maps of wild trees and shrubs and mammals of Europe were very basic, resembling that of Zimmermann (see Camerini 1993b, p. 486), from which Ritter's method was mostly likely derived (see Engelmann 1966, pp. 109–110).

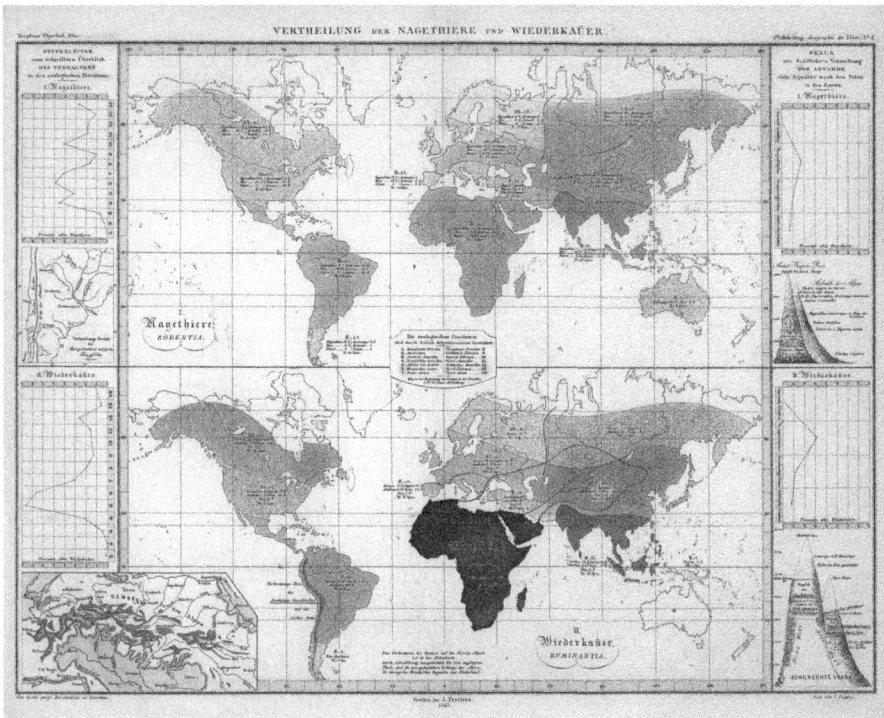

**Fig. 5.3** Berghaus's Division of the Rodents and Ruminants (1845). The original reads, Verteilung der Nagethiere und Wiederkäuer (Source: David Rumsey Map Collection, www.davidrumsey.com)

published in the same year (Fig. 5.3). Berghaus makes the distinction clear in his second edition of 1851:

> Zoological Geography can be viewed from two points of view. One can – 1. pose the question, through which orders, families, [Geschlechter] genera, even [Gattungen] species of the various classes of animals, characterises each of the major land and oceanic regions ["Verteilung"]; or it raises – 2. the question on how animals are distributed in the different zones and regions of the world by their order, [Geschlecht] genus, or [Gattung] species ["Verbreitung"][16] (Berghaus 1851, p. 1, see also Camerini 1993b, p. 502).[17]

---

[16] The original reads "Die Zoologische Geographie lässt sich von zwei Hauptgesichtspunkten betrachten. Sie kann 1. die frage aufwerfen, durch welche Ordnungen, Familien, Geschlechter, ja Gattungen der verschiedenen Tierklassen ein jeder der grösseren Abschnitte der Landfläche und des Ocean charakterisirt ist; order sie wirft; 2. die Frage auf, wie die Tiere einer jeden Klasse in die verschiedenen Zonen und Regionen der Erde verteilt sind, indem bald eine ganze Ordnung, bald ein einzelnes Geschlecht, oder gar eine einzelne Gattung zur Betrachtung gezogen wird." (Berghaus 1851, p. 1, written in modern German).

[17] The sub-title of the Physical Atlas also refers to the distinction between both Verteilung and Verbreitung: "On the main occurrence of inorganic and organic nature according to their geographical distribution and division as depicted [in maps]". Berghaus called the practise

**Fig. 5.4** *Verbreitungsbezirk und Vertheilungsweise der arten* [Areas of distribution and distribution of species] (Schouw 1823, Taf. V)

Berghaus is clearly distinguishing between distributions, which differ from taxon to taxon and divisions (i.e., regions), which contain these distributions. Berghaus divides the world into two well known geographical regions, the Old and New Worlds, then into geographical regions (e.g., Australia, America, Asia) in which has its own set of taxonomic provinces (e.g., Central Asian Province). The distribution maps, resembling those of Zimmerman (1777), are shown as lines indicating distributional boundaries. The concept of "Verteilung" and "Verbreitung" is similar to that in the plant geographers, such as Schouw and Meyen, in which vegetation distributions are mapped in relation to the larger geographical regions. In fact Berghaus (1839) depicts Schouw's 25 regions and their respective vegetational distributions (Fig. 5.4). Here I wish to return to de Candolle (1820) and Nelson (1978) and the notion of stations and habitations:

---

of *Verteilung* "allgemeine zoologichen Geographie" and the practise of *Verbreitung* "specielle zoologische Geographie" (Berghaus 1851, pp. 1–2). Camerini (1993b, 502) calls this "General" and "Special" zoogeography respectively.

> By the term station I mean the special nature of the locality in which each species customarily grows; and by the term habitation, a general indication of the country wherein the plant is native. The term station relates essentially to climate, to the terrain of a given place; the term habitation relates to geographical, and even geological, circumstances [...] The study of stations is, so to speak, botanical topography; the study of habitations, botanical geography (de Candolle 1820, p. 383; translated in Nelson 1978, p. 280).

The distribution and divisions of Berghaus are clearly analogous to that of de Candolle's stations (botanical topography) and habitations (botanical geography), although de Candolle is not mentioned in the fifth volume of the 1851 phytogeography edition. In chapter 1, I stated that the claim made by Nelson (1978) that "ecological biogeography is the study of stations; historical biogeography, the study of habitations" and that "the terms as used by Candolle, have modern counterparts: ecological and historical biogeography" is intriguing. For example, by 1850, the maps by Berghaus and the work of Schouw, Meyen, Wahlenberg and de Candolle for example, all points to there being a division based on the size and nature of an area. The division is resultant of plant geographers using a vegetation classification based on vegetative forms that a product of their immediate environment. The traditional taxonomists however looked at the larger distribution of species, genera and families, which cover a multitude of different environments. In animal geography isothermal barriers in latitude and altitude accounted for smaller distributions, while regions were based on existing geographical regions (e.g., Old and New Worlds, Australia, North America). In others words, de Candolle's stations are based on measurable small-scale geographical phenomenon (e.g., rainfall, temperature), while habitations were mainly focused on large-scale geological phenomenon (e.g., mountains, oceans).

The study area and study organism essentially drove the area and organism classification. By the 1850s and 1860s the divide between each classification already meant there was a rift between an "ecological" and a "taxonomic" method. While these two approaches conflicted, they avoided any confrontation by keeping their studies focused on size. Someone studying tamarind forms in a subtropical climate would not ever conflict with someone working on the geography of legumes. These two approaches however are discordant when we attempt to form regionalisations as seen in the maps of Berghaus.

Berghaus (1851, pp. 8 & 17) groups two taxonomic groups, the carnivores and rodents, into two sets of unique provinces (Table 5.2). The provinces are arbitrary, based on geographical regions rather than on the distributions of the organisms that inhabit them. A.P. de Candolle's habitation, namely "a general indication of the country wherein the plant is native" does not provide us with any indication of how that habitation was selected. The point Swainson and Prichard made is that a natural habitation will provide the animal geographer with a better understanding of the natural distribution of taxonomic groups. Berghaus clearly missed this point. The cartographer shoe-horned in natural distributions into fixed possibly artificial areas.

**Table 5.2** A comparison of carnivore and rodent provinces of Berghaus (1851, pp. 8 & 17)

| Carnivore provinces | Rodent provinces |
| --- | --- |
| North European | North European |
| Middle European | Middle European |
| South European | South European |
| Asian–European overlap | Asian–European overlap |
| North Asian | North Asian |
| Central Asian | Inner Asian |
| Tropical Asian | Tropical Asian |
| Oceanic | – |
| Australian | Australian |
| North African | African |
| Tropical African | – |
| South African | – |
| Arctic American | Arctic American |
| North American | North American |
| Tropical American | Tropical American |
| South American | South American |

Note how several of the provinces are not comparable (i.e., Tropical African versus African) and, the fact that two similar (possibly the same) areas are given different names (i.e., Central Asian versus Inner Asian). Together these two sets of provinces do not overlap (i.e., Tropical Africa is not equivalent to Africa). Non-overlapping area classifications such as this plague zoogeography, meaning that general animal regions are harder to identify

In addition, Berghaus assigned a percentage of genera that are distributed in each of the regions and provinces,[18] providing a rough diversity quotient.

The pursuit of natural regions in animal geography was hampered in part by the concept of isothermal lines. The *Map of the distribution of marine life, illustrated chiefly by fishes, Molluscs & Radiata; showing also the extent & limits of the Homoiozoic belts* (Forbes 1854)[19] is one such attempt at delimiting large marine

---

[18] Berghaus (1851) states that maps were created in the winter of 1843–1844, however the statistics (percentages of distribution) was added later in November 1850.

[19] There was much support for Forbes and his work on Homoiozoic belts "... Forbes' views, to show how profoundly he was impressed with the belief that geographical and climatal conditions were all-powerful controllers of the migrations animals and plants. Forbes was the reformer of the science of geographical distribution" (Hooker 1881, p. 446).

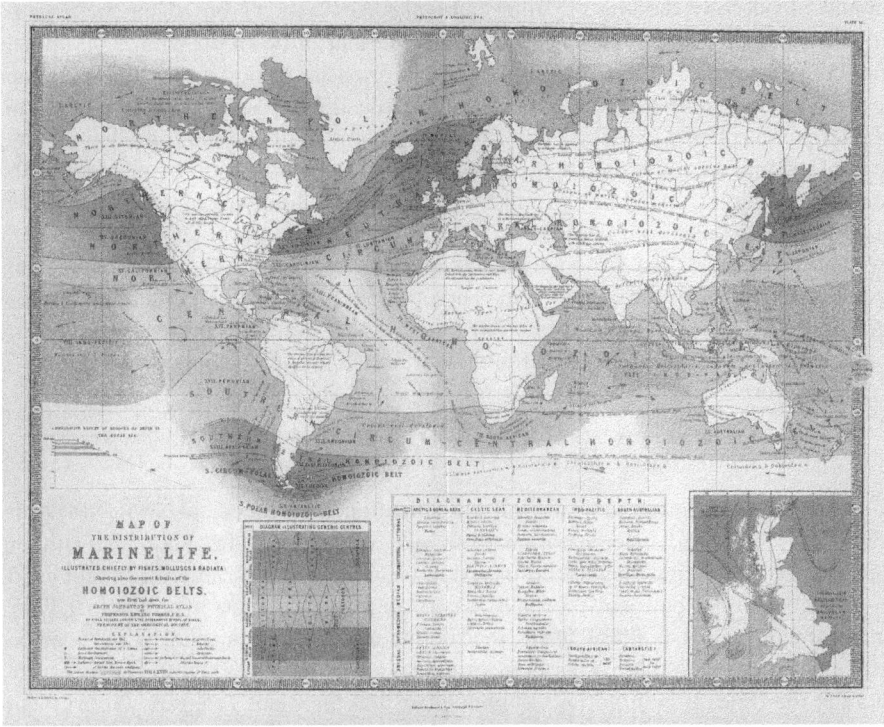

**Fig. 5.5** *Map of the distribution of marine life, illustrated chiefly by fishes, Molluscs & Radiata; showing also the extent & limits of the Homoiozoic belts* of Forbes (1854) (Source: David Rumsey Map Collection, www.davidrumsey.com)

regions using isotherms[20] (Fig. 5.5). But isothermal lines do not account for old distributions, overlap zones and large biogeographical breaks like Wallace's line.[21]

Natural regions were sought at regional levels in animal geography. Schmarda (1853) understood that local environments, vegetation and geography influenced

---

[20] Forbes also included a cross-section "Comparative extent of the regions of depth in the Aegean Sea", which is analogous to the phytogeographical profiles of eighteenth century naturalists.

[21] Interesting to note that in Berghaus (1845), the map titled *Der Indische Archipelagus nach Sal.[Salomon] Müller* actually shows an early line separating the Australasian and Sunda faunas. This predates Wallace's Line (Wallace 1863) by 14 years (see Camerini 1993a, b). Berghaus (1845) points this out clearly: "There is nowhere on the face of the Earth where your find a large difference between animals over a such a short distance as you do in the Indian Archipelago. Although the islands between Java and New Guinea have about the same climate, and many island lie so close to another and are separated by narrow straits, almost every island has its own endemic genera" (Berghaus 1845, p. 24). Interestingly the line drawn by Berghaus is based on the percentages of genera on either side. Unusually, if Berghaus had used this approach in his other maps, he may have drawn natural areas based on distinct biogeographical boundaries. What is not clear is why the division runs straight through Timor, given that each of the islands are treated as separate areas.

animal physiology thereby creating "animal-forms". Areas based on these geographical conditions produced natural areas that had little to do with geopolitical regions (Schmarda 1853, p. 90). These animal forms are analogous to the vegetation forms of the Humboldtians, however Schmarda did not create an animal-form classification scheme to replace the Linnaean system. With animal form delineating natural provinces, the larger regions would be an amalgamation of these animal-forms or of larger taxonomic groups. Schmarda's method was akin to that of the Humboldtians, but to only a degree. There however was a problem. For example, two animal forms would differ the further apart they were from another. Consider an arctic form to that of a tropical form. The problem is that while taxonomic groups changed over space, animal forms did not. New World and Old World vultures, while similar in form belong to different taxonomic groups. Schmarda called these "vicariant forms":

> Vicariant forms are those, which occur in another region, are closely related to those nearby [...] there are vicariant species, genera and even families (Schmarda 1866, p. 425).

With vicariate forms, any natural classification is doomed to failure, as such forms are found in different areas with no connection other than the similar environments in which they occur. By using vicariate forms Schmarda did not need to talk about distributional mechanisms. If animal forms are a product of their environment, and it is these forms that define a region, then there is no need to explain how they got there other than citing the environmental conditions occurring there at the time. It is important to remember that the plant geography movement (i.e., Humboldtians) shunned history[22] as unknowable and Linnaean taxonomy as artificial (Nicolson 1987, 1996). Alternatively, taxonomic zoologists like Phillip Lutley Sclater, who delved into larger taxonomically defined regions, circumvented the need to explain distributions by insisting taxa were created were they were. Unlike vicariate forms, multiple creation was a far more controversial idea, one that needed special pleading (i.e., a Devine creator) rather than to observable natural processes (e.g., climate, temperature, etc.).

---

[22]Dismissing historical processes as valid geographical drivers of modern day distributions did irritate prominent nineteenth century practitioners of plant geography, such as English botanist William Turner Thiselton-Dyer: "At any rate, whatever direction our speculations take, the Australian flora seems to give little support to those who, like Grisebach, ignore the influence of geological change and explain plant-distribution exclusively from the phenomena of climate" (Thiselton-Dyer 1878, pp. 428–429).

# A Multitude of Regionalisations (1858–1899)

*Only the future can lift the veil that still covers some parts of the same (Jordan 1883, p. 217).*[23]

By the middle of the nineteenth century, regionalisation had become a predominant method in plant and animal geography. Most floras, animal studies and geographies had large-scale region divisions, however, many regionalisations of the same taxa did not overlap, creating conflicting area classifications. Given that regionalisation (a.k.a., area classification) is the main methodology used by animal and plant geographers, differences in regionalisation based on similar taxa appear to be artifactual of the methodology. While this seems plausible, it is however, the type of classification used to organise the taxa or vegetational forms that may in fact be driving the result. While we may use the same taxa to determine zoogeographical areas, how they are classified will alter the result. For example, a species that is considered to be part of *Genus* A will alter the distribution if it is considered to be part of *Genus* B in another regionalisation. Moreover, if we classify tree ferns taxonomically we will get a different result if they are classified as tree forms. Given this, it seems implausible that a regionalisation of forms will ever overlap with larger regions, hence Berghaus's "Verteilung" and "Verbreitung" – we either find which organisms are characteristic of a region, or we map the individual distributions. The first denies the possibility of natural regions, as there are implied *a priori*. Berghaus used provinces that were clearly divided into the Old and New Worlds, both geographical not natural biotic regions. Finding natural biotic regions, that is regions that are defined *a posteriori* through understanding the distributions of species, genera and families, would provide a better area classification.

## Plant Regionalisation

By the mid-nineteenth century plant regionalisation was mostly based on the distribution of vegetative forms that were driven in part by temperature, humidity and soils (Unger 1852, pp. 2–4). This differed from the taxonomy-based plant geography (*sensu* de Candolle 1820), which concentrated on larger regions and realms (de Candolle's *habitations*) that were driven by historical processes. Oscar Drude (1890) points this out in the preface to his *Handbook of Plant Geography,* namely that the taxonomy [plant system] is stifling the debate about the principles plant geography and that it would benefit from a "different chain of thought". Drude proposed a sample of seven taxonomic orders as the "touchstone of geographical

---

[23] Original reads "Erst die Zukunft kann den Schleier heben, welcher noch manche Teile derselben bedeckt" Jordan (1883).

botany" to indicate how they link to vegetation forms and where their natural areas occur. However, Drude was careful to distinguish taxonomy from vegetation forms:

> Vegetation forms are biologically conceived and should be separated from the (natural-morphological) botanical taxonomy [Wissenschaft den Pflanzenarten]. By placing these [vegetation forms] into higher units, notions of 'vegetation classes' arise; finally these considerations lead to the implementation of an autonomous biological system (Drude 1890, p. 62).[24]

The "autonomous biological system" that Drude refers to is a separate classification system of vegetative forms. Drude refers to the older Linnaean taxonomy as artificial (i.e., the "so-called natural system") and the new system as "Natural vegetation classification".

In *The Development of Plant Geography in the last hundred years and further and other similar tasks* Adolf Engler divides plant geography in the same way: (1) The floristic-statistical or floristic-systematic which is based on plant taxonomy, which for instance also determines the relationship between of the individual plant families within plant regions and how those regions are related and; (2) Floristic-physiognomic, which is based on vegetation forms. Engler however, proposes another category; (3) Floristic-geographical, "based on the facts in both 1 and 2, attempts to subdivide the Earth, its continents or smaller areas, based on their plant communities" (Engler 1899, p. 13). What Engler means is that the combination of both the floristic-physiognomic and floristic-systematic approaches would work toward a third approach. While Engler does not use the term "unity", his argument certainly smacks of a unified approach of a field that had been divided in its use of classification and what it considered to be natural. Engler attempted to smooth over this theoretical crack by pointing out epistemological similarities between both approaches, namely regionalisation. By 1899, Engler was attempting to unify plant geography, while animal geography was slowly dividing into two different approaches.

## *Animal Regionalisation*

Animal regionalisation began in earnest with the regionalisation of Sclater (Fig. 5.6). The areas of Schmarda had to all intents and purposes disappeared from the literature. Animal geographers were staunch taxonomists and areas based on vicariant forms may have been unpalatable and certainly of no use to those who only used Linnaean taxonomy. Rather the conflict between late nineteenth

---

[24]The original reads: "Die Vegetationsformen sind biologisch aufzufassen und vom eigentlichen (natürlich-morphologischen) System des Pflanzenreiches, welches zugleich den von der Wissenschaft den Pflanzenarten zuerteilen Namen anzeigt, getrennt zu halten. Indem man sie wiederum zu höheren Einheiten vereinigt, kommt man zum Begriffe der 'Vegetationsklassen'; es führen also diese Betrachtungen schliesslich zur Aufstellung eines eigenen, biologischen Systems" (Drude 1890, p. 62).

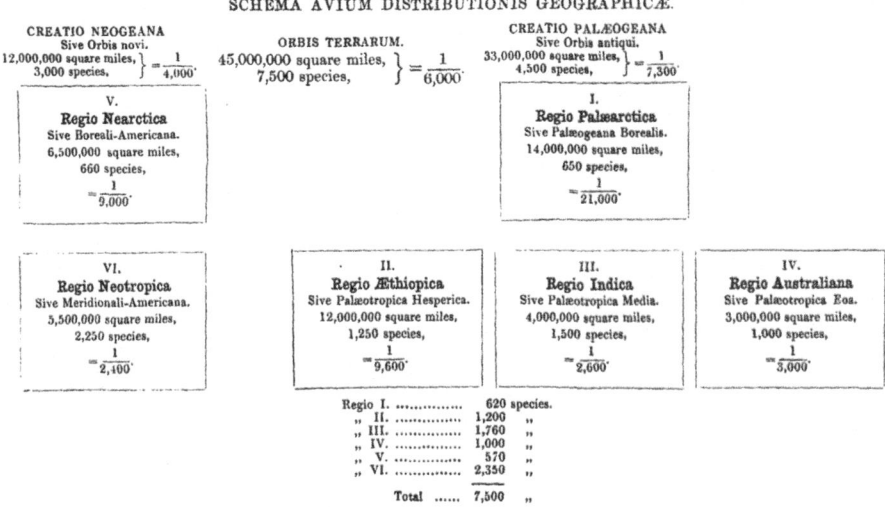

**Fig. 5.6** *Schematic Diagram of the Geographic Distribution of Birds* (Sclater 1858, p. 145). "Philip L. Sclater's division of the terrestrial world into six faunal regions based on the distribution of birds. The approximate area and number of species living in each region was used to estimate the area occupied by a single species. The Neotropics, region VI, is the densest, with one species per 2,400 mile$^2$. The list at the *bottom* of the figure gives alternate numbers of species per region" (Parenti and Ebach 2009, p. 28)

century regionalisations of animal geographers was along the lines of practicality versus precision. Most late nineteenth century animal geographers denied they had enough convincing information to accurately divide the world into large equal regions. Practically minded animal geographers, however, adopted large divisions on the proviso that future studies would refine these areas. The debate flared up in the scientific literature and seemingly never resolved itself.[25] For example, an

---

[25] Scientists seemingly despair of the period after 1880: "The late 1800's are often considered a rather sterile period in the history of biogeography because workers were engaged in debates about the borders of regions, a descriptive endeavour [...] Actual analyses of biogeographical concepts in the 1880's appear to be lacking, however. Thus von Hofsten [1916] and Schmidt (1955) remained silent on that period" (Vuilleumier 1988, p. 19). Vuilleumier's view (similar to that of Nelson 1978) was that "biogeography after was largely influenced by Darwin's work, but then so was all of biological thought" (Vuilleumier 1988, p. 19). Compare the view of Vuilleumier (1988) with that of Good (1955): "The work of Drude, however, is more generally known from his *Handhuch der Pflanzengeographie*, which appeared in 1890 and contained among other things an improved floristic classification. This, however, though an important book, said little that was entirely new and gives the impression rather of belonging to the end of an epoch. [...] Thus it would seem that by the eighteen-nineties the forward urge provided by Darwinism had begun to work itself out and that some new impulse was due. This came in the form of a concentration upon the relation between the plant and its immediate environment or habitat, a new approach or point of view to which was given the name 'plant ecology,' or 'oecology,' as it was first spelled. The first

anonymous letter titled *The Nearctic Region and its Mammals*, published in *Natural Science* in 1893, was a reply to two articles penned by Merriam (1892) and Allen (1892) on new biogeographic regions of North America:

> It would appear that the Munroe doctrine of 'America for the Americans,' is little heeded by the biologists of that Continent. Although the best European authorities on the geographical distribution of animals have long ago conceded to the northern half of the New World the rank of one of the six primary divisions of the earth's surface, under the name of the 'Nearctic Region', our American friends will have none of it (Anon. 1893, p. 288).

The anonymous letter decried the use of different regions to define mammal distribution of North America. Rather than using the widely accepted Nearctic, Merriam and Allen chose instead to create new areas (Fig. 5.7). Allen used "Boreal" and "North American Region", while Merriam (following Cope) used the term "life-zones" that were part of the "Sonoran region". Together Allen and Merriam divided up the North American region into three "constituent parts, and in repudiating the view of Sclater and Wallace that it should form one of the main zoo-geographical divisions of the earth's surface" (Anonymous 1893, p. 288). Allen and Merriam were interested in precision and not convenience:

> Wallace, in writing of the principles of which Zoological regions should be formed, expresses the opinion that 'convenience, intelligibility, and custom', should largely guide us'. But I quite agree with America's most distinguished and philosophic writer on distribution, Dr. J.A. Allen, that in marking off the life regions and subregions of the earth, truth should not be sacrificed to convenience (Merriam 1892, p. 64).

Both Allen and Merriam were professors of biology and, as professionals, would never allow convenience to get in the way of detailed study. Wallace on the other hand, was an amateur who wanted a general idea of geographical distribution of animals. The detail can be seen in comparing Merriam (1892) with Wallace (1876a, b). Merriam listed all known works regarding the Nearctic region, sub-region by sub-region and compared them in order to calculate the boundaries of his life zones. The detail of the work is incredible. Merriam's 65 page work is far more rigorous and detailed than Wallace's 39 page chapter on the Nearctic (Wallace 1876a, vol. 2, pp. 114–153) and Sclater's even smaller contribution. Given this, why the anonymous letter?

Most likely because Wallace's regions were based on those of Sclater (1858), and it is Sclater's areas, not Wallace's, that Allen critiques:

> These divisions, as has been urged recently in the favour [by Wallace (1876a, b)], are *convenient* and *easy to remember*, since they are approximately equal in size, are easily defined, and avoid complicated boundaries (Allen 1892, pp. 211–212, original italics).

Allen's issue was two fold: why divide the northern areas into Old World and New world this entirely "ignoring the close similarity of life throughout the cold temperate and arctic regions of the globe" (Allen 1892, p. 211) and, why ignore

---

principles of this new discipline, which, as we shall see, has since become the sister of the older plant geography in the stricter sense …" (Good 1955, p. 751, original italics).

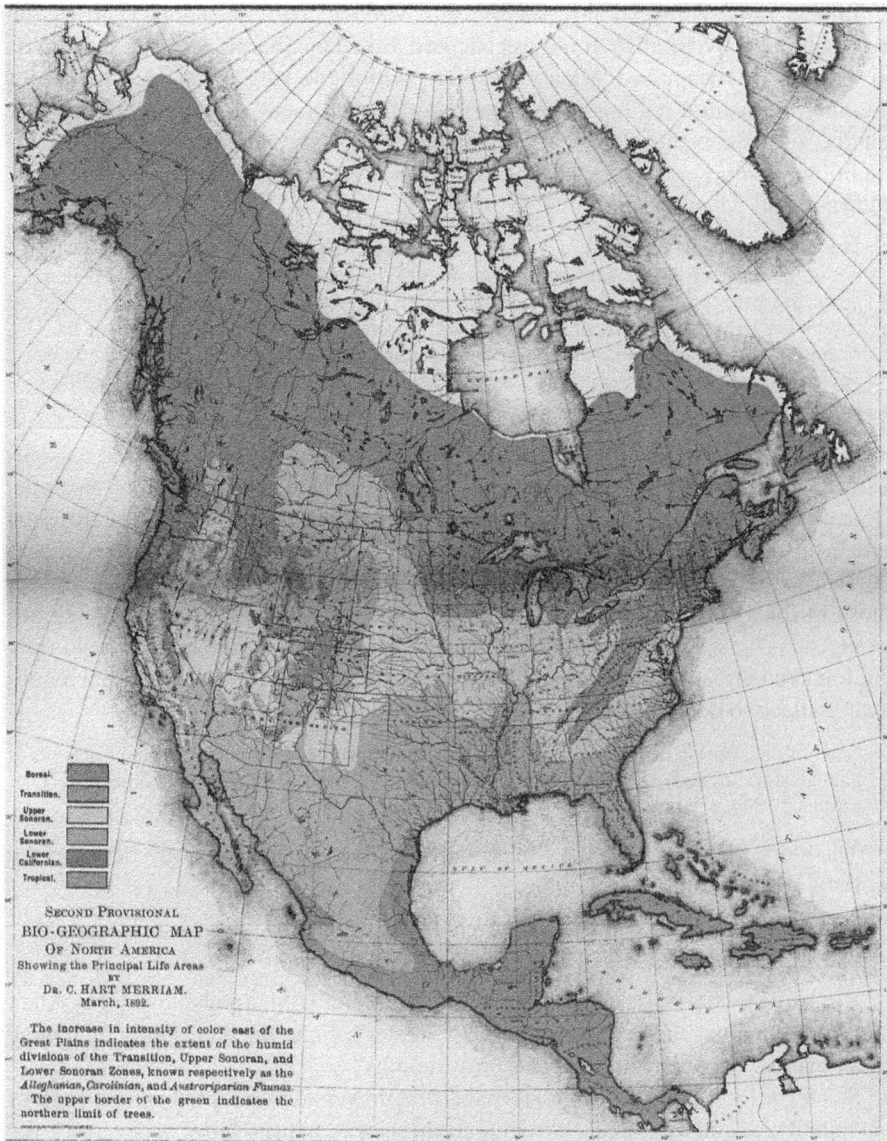

**Fig. 5.7** *Second Provisional Bio-Geographic Map of North America showing the principal Life Areas* (Merriam 1892) (Image courtesy of Erin Clements Rushing and the Smithsonian Institution Libraries)

new improved schemes by Huxley (1868) and Günther (1858) when many North American and continental zoogeographers have ignored the term Nearctic?[26]

---

[26]"We reject the term 'Nearctic' proposed by Mr. P. L. Sclater, and adopted by Mr. A. R. Wallace, for America north of Central America, for the reason that it seems to us an unnatural and artificial

In the same third volume of *Natural Science*, there is another article on zoogeography by English zoologist Richard Bowdler Sharpe. In it Sharpe refers to the both Merriam and Allen's work as "admirable [...] with excellent maps defining the natural regions of the North American fauna" (Sharpe 1893, p. 100). Sharpe's article is an attempt at defining the world's avifaunal regions using the traditional realms and regions proposed Wallace. However, Sharpe is dismissive of Merriam and Allen's "admirable" work:

> I cannot understand why the word 'Nearctic' should be discarded. It was given by Dr. Sclater not in the sense of 'arctic' but 'northern' region of the New World, and is, in my opinion, apart from the priority which commands respect for its retention, a most simple and expressive term. My American colleagues will understand that if I have not carried their system of nomenclature unto the zoo-geographical regions of the Old World, it is not from any want of respect to their work, for I heartily agree with their conclusions as regards North America (Sharpe 1893, p. 101).

Sharpe is clearly retaining Wallace's regions because of nomenclatural reasons. Already North America had several terms associated with it, including Nearctic, Boreal, Sonoran etc. Clearly the laws of distribution were different for different organisms.[27] Some organisms were more influenced by climate than others, meaning that a suite of zoogeographical regions for mammals may indeed be different for those for birds or beetles. The more these studied prevailed the more new regions would be proposed and a plethora of names would enter the literature would only confound an already confusing field. By 1894, the proliferation of areas and names lead Wallace to declare:

> Laws of distribution can only be arrived at by comparative study of the different groups of animals, for this study we require a common system of regions and a common nomenclature (Wallace 1894, p. 612).

Theodore Gill also saw laws of distribution the contending issue between the work of Wallace and Allen,

> On a comparison of the respective schemes of Messrs. Wallace and Allen, it is obvious that they must have been influenced by quite different considerations (Gill 1885, p. 5).

But names were only a small part of the problem. If Sharpe believed that Allen and Merriam areas that were analogous to Sclater's Nearctic, then surely there may be an overlap between mammal and birds regions?

Sclater's 1858 areas were possibly the first universally accepted natural regions in animal geography. Sclater's regionalisation was a firm move away from the

---

term [...] It is to be hoped that the term will not be adopted by American writers, as it is not by German and French writers, and we heartily endorse Mr. J. A. Allen's protest against the use of the term by American writers on this subject" (Packard 1883, p. 363, also in Allen 1892, p. 212, footnote 1).

[27]Theodore Gill (1885) provides great summary of the differences between the areas of Wallace and Allen. For the most part, Allen used isothermal lines, altitude and other geographical measurements to determine areas, hence the detail. Wallace merely focused on certain mammal distribution that he thought were important in determining regions.

geographical isothermal lines of Forbes and Merriam and the physiognomy of Schmarda's distribution of animal forms. The Sclaterian regions also avoided the trivialities of individual taxic distributions at a regional level as seen in the work of Allen and Merriam. What made Sclater's regions appealing is that they were convenient and were usable by a far greater portion of the animal geographical community. That is not to say that the life-zones of Merriam, for example, were useless. Rather, life-zones were too specific, too detailed and rigid to make the board-brush comparisons between regions through time. For such comparisons you need 'convenience, intelligibility, and custom', perhaps explaining why all six Sclaterian regions are still used in twenty-first century biogeography.[28]

The popularity of the Sclaterian regions only came later in the twentieth century. During the late nineteenth century, American zoogeographers such as Allen and Merriam considered these areas artificial even though they were "convenient" and "easy to use". Why then were the Sclaterian regions adopted when, according to Allen (1892), there were better regions to use?

Zoogeography is a complicated field, with very little data and much speculation. Few regionalisations were practical or easy to use. The regions of Allen and Merriam, still popular by the 1940s (see Dice 1943) were only adopted at the smaller regional level (North America). Then there were taxic discrepancies between subregions and regions. Günther (1858) for example, thought that there were enough differences between Sclater's ornithological and his own herpetological faunas to justify two separate sets of zoogeographical regions. In wanting to be detailed and precise, zoogeographical regionalisation focused on single taxa, such as snakes, rather than larger taxa such as vertebrates, leading zoologists such as Günther to give an "account of the geographical distribution of those animals, to the knowledge of which especially I have latterly devoted myself; and often referring to that paper, I shall show how far I can agree with the general views contained therein, and whether these parts of the natural kingdom give us a division of the earth's surface into the same natural provinces" (Günther 1858, p. 373).

In effect what Günther is saying is that only a specialist with a complete knowledge of their group is in a position to provide divisions of the Earth. Sclater did it with birds, therefore Günther could do it with snakes and toads. What Sclater and Günther, and later Allen, Merriam and Packard did not realise is that regionalisation is more about generalisation than it is about precision. It is seemingly impossible to account for every genus and species distribution within group into to determine a broad generalisation. For example, it would be the same as dividing up North America or the Boreal region, using a single taxon. Broad geographical patterns are found by broadly looking at many unrelated taxa. The failure of nineteenth century regionalisation was that every taxon was assigned with its own unique set of areas, leaving generations of zoogeographers with either conflicting regionalisations or similar regions with a plethora of new names. Allen's argument can be countered:

---

[28]The regions of Allen (1892) were redundant by the mid twentieth century. Merriam's life-zones however are still in use in the twenty-first century ecology.

these divisions are *convenient* and *easy* to remember, *because* they are easily defined *and* avoid complicated boundaries.

Wallace was possibly the only zoogeographer to consider the merits of a convenient and easy to use regionalisation an asset to zoogeographers. Like Schmarda, Wallace created a regionalisation for all land vertebrates, but unlike Schmarda, Wallace used the distributions of families and genera (not species) rather than of variant forms. Wallace summarises his method in the preface of his first volume of *The Geographical Distribution of Animals*:

> [To establish the science of distribution], uniformity of treatment appeared to me essential, both as a matter of principle, and to avoid all imputation of partial selection of facts, which may be made to prove anything. I determined, therefore, to take in succession every weep-established family of terrestrial vertebrates, and to give an account of the distribution of all its component genera, as far as the materials were amiable. Species, as such, were systematically disregarded, – firstly, because they are so numerous as to be unmanageable; and, secondly, because they represent the most recent modifications of form, and are therefore not so clearly connected with geographical changes as are the natural groups of species termed genera; which may be considered to represent the average and more permanent distribution of an organic type, and to be more clearly influenced by the various known or inferred changes in the organic and physical environment (Wallace 1876a, pp. vii–viii).

Compare Wallace's method with that of Schmarda,

> The area that a species [Gattung], genus [Geschelcht] or family covers, we will call the area of distribution of a species [Gattung], genus [Geschelcht] or family. The area of distribution is made up of different distributions, in some animal forms it is large and in others very small (Schmarda 1853, vol. 1, p. 63).

Schmarda is concerned with animal forms and their distributions within areas defined by higher taxonomic groups. Animal forms however, are ahistorical as they are based on the physiological characteristics found within a particular present day environment. Wallace would not consider such forms as relevant, as he is interested in the historical aspects of an area. Therefore only genera and families would do, not their physiological animal forms.

By moving away from single taxa and ahistorical (or present day) physiological forms that are characteristics of their environment, to a more general distribution of higher taxa within a historical areas, Wallace created large regions that could be used by across zoogeography. However, by ignoring the animal forms (adaptations to certain environments) and species distributions, Wallace found himself at odds with the zoogeographical community as to the Nearctic region,

> When I worked many years ago on this subject, I doubted much whether the now-called Palearctic and Nearctic regions ought to be separated (Darwin to Wallace, June 5, 1876, in Marchant 1916, p. 286).

> ... [I] am in a terrible state of doubt & am fast becoming a disbeliever in the Nearctic Region! but I hope I may find some decent excuse for remaining in the orthodox faith (Newton to Wallace, April 28 1875, Wallace Letters Online).

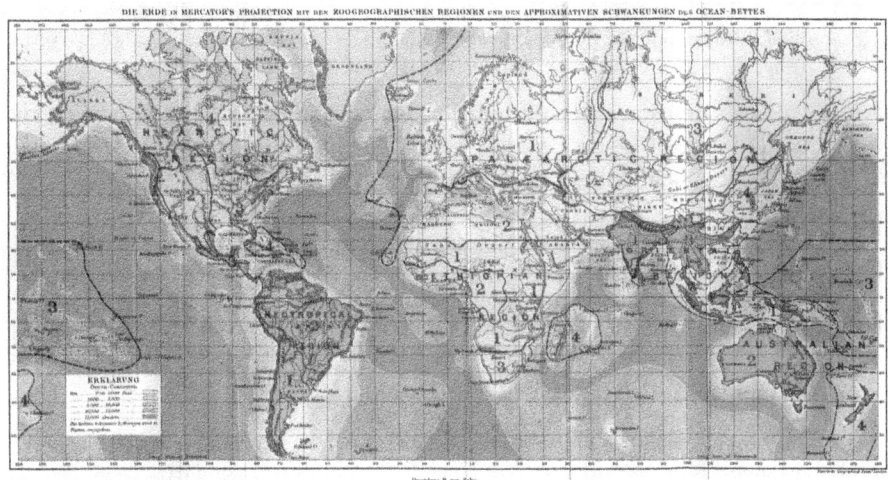

**Fig. 5.8** *The Geographical Distribution of Animals* (Wallace 1876b). Note this is the 1876 German edition (Source: http://upload.wikimedia.org/wikipedia/commons/2/2c/Wallace03.jpg)

> Now taking the distribution of these genera from Baird & Sclater I find there are 13 genera wholly confined to the Nearctic Region; 20 more of which all the species are Nearctic, but some of them extend to Mexico [,] a few more to Guatemala … (Wallace to Newton, May 8, 1875, Wallace Letters Online).

> I am sorry you are so disturbed about distinctness of Nearctic & Neotrop[ic] regions. Your statistics do not in the slightest degree affect my conviction that they sh[oul]d be kept absolutely distinct (Wallace to Newton, May 26, 1875, Wallace Letters Online).

> I will only remark now that you proceed on your supposition that my "Holarctic" Region = your Palaearctic and Nearctic – whereas the southern boundaries of this last are, in my opinion and that of several American zoologists, very uncertain [ … ] Thus a very considerable number of the genera, which you assign to your Nearctic and Palaearctic Regions, belong really to more southern areas, and by their elimination your lists would present a very different aspect. Again too, you have omitted from your Nearctic list all the Palearctic genera of birds which inhabit Alaska, and if I am not mistaken these are several Mammals also, making Alaska essentially Palaearctic (Newton to Wallace, June 17 1875, Wallace Letters Online).

Wallace' adherence to the Sclaterian system is admirable. Wallace knew that Sclater's system did differentiate between the most northern regions, even though they did share several taxa (as noted by Zimmermann's classification of *Quadrupeds of both the Old and New World* in his 1777 *Specimen*). Wallace wanted to break the mould of nineteenth century zoogeography (Fig. 5.8):

> The divisions in use till quite recently were of two kinds; either those ready made by geographers, more especially the quarters or continents of the globe; or those determined by climate and marked out by certain parallels of latitude of by isothermal lines. Either of these methods was better than none at all; [but] it will be evident, that such divisions must have often been very unnatural, and have disguised many of the most important and interesting phenomena which a study of the distribution of animals presents to us [ … ] The merit of

initiating a more natural system, that of determining zoological regions, not by any arbitrary or *a priori* consideration but by studying the actual ranges of the more important groups of animals, is due to Mr. Sclater ... (Wallace 1876a, b, vol. 1, pp. 52–53).

But Wallace's division appeared archaic, distinguishing the Old and New Worlds, which Sclater termed *Creatio Paleogeana* and *Creatio Neogeana* respectively and, proved to be unpopular with some zoogeographers:

> The six zoological regions laid down by Mr. Sclater, and so admirably sketched out by Mr. Wallace, have been so very generally accepted by naturalists that it may be considered as almost presumptuous for any one to attempt at this late hour a revision of the same. But yet the evidence concerning the position of at least one of these – the Nearctic – is in many respects so negative-indeed, it might be said so directly contradictory that a reconsideration is rendered almost imperative (Heilprin 1882, p. 316).

The debate worked its way into the pages of *Nature*,

> [Heilprin] seeks to show that the Neoarctic [sic] and Palearctic should form one region, for which he proposes the somewhat awkward name 'Triarctic Region', or the region of three northern continents (Wallace 1883, p. 482).

> Briefly stated, it is maintained that [...] the Neoarctic [sic] and the Palaeoarctic [sic] faunas taken individually exhibit, in comparison with the other regional faunas (at least the Neotropical, Ethiopian, and Australian), a marked absence of *positive* distinguishing characters, a deficiency which in the mammal extends to families, genera, and species, and one which, in the case of the Neoarctic [sic] region, also equally (or nearly so) distinguishes the reptilian and amphibian faunas (Heilprin 1883, p. 605, original italics).

> The facts of zoogeography are so involved, and often apparently contradictory, that a skilful dialectician with the requisite knowledge can make plausible argument for antithetical postulates. Prof. Heilprin, being a skilful dialectician and well informed, has submitted a pretty argument in favour of the union of the North American or 'Neacrtic' and Eurasiatic or 'Palaearctic' (Gill 1883, p. 124).

The issue here is methodology, namely, how do you distinguish one region from another. Considering that Heilprin and Wallace have two different methods (Heilprin acknowledges species distributions whereas Wallace does not), it is not surprising they designate areas in different ways. Heilprin's insistence that Wallace is basing his regions on negative characteristics, that is absence of taxa, as distinguishing characteristics, is unfounded. Comparative biology is based on relating what two taxa or faunas share and how they differ (i.e., positive and negative characteristics). While Heilprin also suggests using positive and negative characteristics to compare different faunas, he sees the close relationship between the Nearctic and Palaearctic as a reason to unite them as a single area, rather than treat them as two separate closely related areas. The idea of a relationship between regions is an often-explored concept in statistical zoogeography. However, Sclater was possibly the first to suggest a systematic way to compare areas,

> ... but little or no attention is given to the fact that two or more of these geographical divisions may have much closer relations to each other than to any third, and, due regard being paid to the general aspect of their Zoology and Botany, only form one natural province or kingdom (as it may perhaps be termed), equivalent in value to that third (Sclater 1858, p. 131).

Such close relationships, however, would only be seen as a way to combine regions.

For Sclater and Wallace, the combination of the Palaearctic and Nearctic (i.e., Holarctic) would conflict with the higher classification of the Palaeogeana and Neogeana.[29]

The debate about Sclaterian regions, their delimitation and extent, was stymied by a new generation of zoogeographers such as Ortmann, who stated:

> It is incorrect to regard the creation of a scheme [of regions] of animal distribution as an important feature or purpose of zoogeographical research. Thus we are justified in saying that zoogeographical study, as introduced by Wallace [and Sclater], is not directed in the proper channels [and results in] fruitless discussions on the limits of the zoogeographical regions (Ortmann 1902, after Heads 2005, p. 87, in Williams and Ebach 2008, p. 236).

The descriptive nature of zoogeography had come to pass and, like Alphonse de Candolle's dismissal of his father's plant geography as an artificial system, was seen as a detriment to science "when they considered to be natural" (de Candolle 1855, pp. 1304–1305). Zoogeography had reached a similar stage to that in phytogeography in the mid 1850s: the study of taxonomic distribution and fixed zoogeographical regions was subsumed with a more, seemingly, dynamic study of fluid fauna (Williams and Ebach 2008; Williams 2007). Here we return back to Mayr's statement, "... zoogeography has had a similar fate very much like taxonomy. It was flourishing during the descriptive period of biological sciences. Its prestige, however, declined rapidly" (Mayr 1944, p. 1). Mayr truly dismantled zoogeographical regionalisation:

> Eventually it was realised that the whole method of approach – *Fragestellung* [question] – of this essentially static zoogeography was wrong. Instead of thinking of fixed regions, it is necessary to think of fluid faunas ... (Mayr 1946, p. 5, in Williams and Ebach 2008, p. 237).

By the beginning of the twentieth century, zoogeographical regionalisation was in decline. Phillip Lutley Sclater and his son William Lutley Sclater, published *The Geography of Mammals* (Sclater and Sclater 1899), in which mammals are divided up into the six Sclaterian regions. Reading the introduction to Sclater and Sclater (1899), one would never know that zoogeographical regionalisation was slowly coming to an end:

> Let us, therefore, dismiss from our minds for the moment the ordinary notions of physical and political geography, and consider how the earth's surface may be naturally divided into Primary Regions, taking the amount of similarity and dissimilarity of animal life as our sole guide (Sclater and Sclater 1899, pp. 1–2).

The popularity vegetation and their regions was that they were quantifiable, through the measurement of climatic factors such as rainfall and temperature as well as geographical factors like soils and so on. While a similar method could be

---

[29] Merriam (1892) and Theodore Arldt (1906) both contain tables comparing the zoogeographical classifications of mid to late nineteenth century and early twentieth century workers.

employed for marine zoogeographical regions (e.g., Homoiozoic belts), terrestrial zoogeographical regions could only be found through taxonomic distributions. In other words, regions based on vegetation were easier to quantify than regions based on taxonomic distributions. Perhaps this is why taxonomic regions survived in Wallace's revised Sclaterian regions. The generation that included the likes of Ortmann (1908) and Cowles (1908) had shunned taxonomic practise in plant and animal geography as rigid and old fashioned.[30] New more rigorous methods, like those used to find vegetation, needed to be adopted in both taxonomy and biogeography. Ortmann, like plant ecologists, sought to understand the geographical and physiological processes responsible for zoogeographical distributions and habitats – ecological zoology had begun (see Nyhart 2009, p. 338–339).

# References

Agassiz, L. (1854). Sketch of the natural provinces of the animal world and their relations to the different types of man. In Nott, J. C., & Gliddon, G. R. (eds.); additional contributions from Agassiz, L., Usher, W., and Patterson, H.S. *Types of mankind: Or, ethnological researches, based upon the ancient monuments, paintings, sculptures, and crania of races, and upon their natural, geographical, philological, and biblical history: Illustrated by selections from the inedited [sic] Papers of Samuel George Morton, M.D* (pp. lviii–lxxvii). Philadelphia: Lippincott, Grambo and Co.

Allen, J. A. (1892). The geographical distribution of North American mammals. *Bulletin of the American Museum of Natural History, 4*, 199–243.

Anonymous. (1893). The Nearctic region and its mammals. *Natural Science, 3*, 288–292.

Appel, J. W. (1994). Francisco José de Caldas: A scientist at work in Nueva Granada. *Transactions of the American Philosophical Society, New Series, 84*, 1–154.

Arldt, T. (1906). Die tiergeographischen Reiche und Regionen. *Geographische Zeitschrift, 12*(4), 212–222.

Berghaus, H. (1838). *Physikalischer Atlas*. Abteilung 5: Pflanzengeographie. Gotha: Justus Perthes.

Berghaus, H. (1839). *Physikalischer Atlas. Abteilung 5: Pflanzengeographie*. Gotha: Justus Perthes.

Berghaus, H. (1845). *Physikalischer Atlas. Abteilung 6: Zoologische Geographie*. Gotha: Justus Perthes.

Berghaus, H. (1851). *Physikalischer Atlas*. Abteilung 6: Zoologische Geographie. Gotha: Justus Perthes.

Bourguet, M.-N. (2002). Landscape with numbers: Natural history, travel and instruments in the late eighteenth and early nineteenth centuries. In M.-N. Bourguet, C. Licoppe, & H. O. Sibum (Eds.), *Instruments, travel and science. Itineraries of precision from seventeenth to the twentieth century* (pp. 96–125). New York: Routledge.

---

[30]Cowles considered taxonomic practises outside of ecology to be artificial: "Taxonomy must be scientific. It must require for its devotees a training as rigid as that required by professional workers in morphology, physiology or ecology. Species-making by taxonomic tyros must be abandoned [...] These things will not, be endured much longer; a little more and the sinning taxonomists will be cast out into the outer darkness where there shall be wailing and gnashing of teeth" (Cowles 1908, pp. 270–271).

Bowen, M. (1983). *Empiricism and geographical thought. From Francis Bacon to Alexander von Humboldt.* Cambridge: Cambridge University Press.
Camerini, J. (1993a). Evolution, biogeography, and maps: An early history of Wallace's line. *Isis, 84,* 700–727.
Camerini, J. (1993b). The physical atlas of Heinrich Berghaus: Distribution maps as scientific knowledge. In R. G. Mazzolin (Ed.), *Non-verbal communication in science prior to 1900* (pp. 479–512). Florence: Leo S. Olschki.
Cowles, H. C. (1908). An ecological aspect of the conception of species. *The American Naturalist, 42,* 265–271.
de Buffon, G. L. L. C. (1761). *Histoire naturelle: générale et particulière, servant de suite á l'histoire des animaux quadrupèdes* (Vol. 9). Paris: L'Imprimerie Royale.
de Caldas, F. J. (1802). *The Imbabura volcano in Ecuador, showing the spatial location of 30 native plant species and their elevation.* Biblioteca Ecuatoriana Aurelio Espinoza Polit, Cotocoyao, Quito, Ecuador (unpublished).
de Candolle, A. P. (1805). Explication de la carte Botanique de la France. In J. B. P. A. de M. de Lamarck & A. P. de Candolle (Eds.), *Flore française, ou descriptions succinctes de toutes les plantes qui croissent naturellement en France, disposées selon une nouvelle méthode d'analyse, et précédées par un exposé des principes élémentaires de la botanique* (3rd ed.). Paris: Desray.
de Candolle, A. P. (1813). *Théorie Élémentaire de la Botanique, ou Exposition des Principes de la Classification Naturelle et de l'Art de Décrire et d'Etudier les Végétaux.* Paris: Déterville.
de Candolle, A. P. (1817). Mémoire sur la géographie des plantes de France, considérée dans ses rapports avec la hauteur absolue. *Mémoires de Physique et de Chimie de la Société d'Arcueil, 3,* 262–298.
de Candolle, A. P. (1820). Essai élémentaire de géographie botanique. In *Dictionnaire des Sciences Naturelles* (Vol. 18, pp. 1–64). Paris: F. Levrault.
de Candolle, A. L. P. P. (1855). *Géographie botanique raisonnée.* Paris: Masson.
Dice, L. R. (1943). *The biotic provinces of North America.* Ann Arbor: University of Michigan Press.
Drude, O. (1890). *Handbuch der Pflanzengeographie.* Stuttgart: J. Engelhorn Verlag.
Dupuis, C. (1974). Pierre André Latreille (1762–1833): The foremost entomologist of his time. *Annual Review of Entomology, 19,* 1–14.
Ebach, M. C., & Goujet, D. (2006). The first biogeographical map. *Journal of Biogeography, 33,* 761–769.
Engelmann, G. (1966). Carl Ritter 'Sech Karten von Europa' Mit einer Abbildung. *Erdkunde, 10,* 104–110.
Engler, A. (1899). *Die Entwicklung der Pflanzengeographie in den letzten hundert Jahren und weitere Aufgaben derselben.* Berlin: Humboldt-Centenar-Schrift der Gesellschaft für Erdkunde zu Berlin, Kühl.
Fabricius, J. C. (1778). *Philosophia Entomologica.* Hamburg: Impensis Carol. Ernest. Bohnii.
Feuerstein-Herz, P. (2004). *Eberhard August Wilhelm von Zimmermann (1743–1815) und die Tiergeographie.* PhD thesis, Technischen Universität Carolo-Wilhelmina zu Braunschweig, Braunschweig, pp. 1–389.
Forbes, E. (1846). On the connexion between the distribution of the existing fauna and flora of the British Isles and the geological changes which have affected their area, especially during the epoch of the Northern Drift. *Memoirs of the Geological Survey of Great Britain, 1,* 336–432.
Forbes, E. (1854). [1856] Map of the distribution of marine life, illustrated chiefly by fishes, molluscs and radiata; showing also the extent and limits of the homoiozoic belts. In A. K. Johnston (Ed.), *The physical atlas of natural phenomena* (Plate 31). Edinburgh: William Blackwood and Sons.
Forster, J. R. (1778). *Observations made during a voyage round the world, on physical geography, natural history, and ethic philosophy.* London: G. Robinson.
Gill, T. (1883). The northern zoogeographical regions. *Nature, 28,* 124.
Gill, T. (1885). The principles of zoogeography. *Proceedings of the Biological Society of Washington, 2,* 1–39.

Giraud Soulavie, J.-L. (1780). *Géographie de la nature, ou distribution naturelle des trois règnes sur la surface de la terre. Suivie de la Carte Minéralogique, Botanique, & c. du Vivarais où cette distribution naturelle est représentée*. Ouvrage qui sert de préliminaire à l'Histoire Naturelle de la France Méridionale, &c. dont on va publier les deux premiers Volumes & à l'Histoire Ancienne & Physique du Globe Terrestre. Hôtel de Venise, Cloître Saint-Benoît. Et chez le Sieur Dupain-Triel, Ingénieur-Géographe du Roi, rue des Noyers. MDCCLXXX, Paris.

Giraud Soulavie, J.-L. (1783a). *Histoire naturelle de la France méridionale. Seconde Partie. Les Végétable. Tome I. Contenant les principes de la Géographie physique du règne végétal, l'exposition des climats des Plantes avec des Cartes pour en exprimer les limites*. Belin, rue Saint-Jacques, Paris.

Giraud Soulavie, J.-L. (1783b). *Histoire naturelle de la France meridionale* (Vol. 1). Paris: Chez J.F. Quillau, Mérigot l'aîné, Belin.

Good, R. (1955). Plant geography. In: E. B. Babcock, J. W. Durham, & G. S. Myers (Eds.), *A century of progress in the natural sciences 1853–1953*. Published in Celebration of the Centennial of the California Academy of Sciences (pp. 747–765). San Francisco.

Günther, A. (1858). On the geographical distribution of reptiles. *Proceedings of the Zoological Society of London, 26*, 373–398.

Güttler, N. (2015). Drawing the line: Mapping cultivated plants and seeing nature in nineteenth-century plant geography. In D. Phillips & S. Kinsland (Eds.), *New perspectives on the history of life sciences and agriculture* (Archimedes, Vol. 40, pp. 27–52). New York: Springer.

Heads, M. J. (2005). The history and philosophy of panbiogeography. In J. Llorente & J. J. Morrone (Eds.), *Regionalización Biogeográfica en Iberoamérica y Tópicos Afines* (pp. 67–123). México City: Universidad Nacional Autónoma de México.

Heilprin, A. (1882). On the value of the "Nearctic" as one of the primary zoological regions. *Proceedings of the Academy of Natural Sciences of Philadelphia, 34*, 316–334.

Heilprin, A. (1883). On the value of the 'Neoarctic'[sic] as one of the primary zoological regions. *Nature, 27*, 606.

Hooker, J. D. (1844–1859). *Flora Antarctica: The botany of the Antarctic voyage* (3 vols.). London: Reeve Brothers.

Hooker, J. D. (1881). Presidential address to the Geographical Section of the British Association – On geographical distribution. British Association Address Reports 727–738; *Nature, 24*, 443–448.

Huxley, T. H. (1868). On the classification and distribution of the Alectoromorphae and Heteromorphae. *Proceedings of the Zoological Society of London, 36*, 294–319.

Illiger, J. K. W. (1815). Überblick der Säugthiere nach ihrer Vertheilung über die Welttheile. *Abhandlungen der physikalische Klasse der Koeniglich-Preussischen Akademie der Wissenschaften, 1804–1811*, 39–159.

Jackson, S. T. (2009). Introduction: Humboldt, ecology, and the *Cosmos*. In A. von Humboldt & A. Bonpland (Ed.), *Essay on the geography of plants* (edited by S. T. Jackson) (pp. 1–52). Chicago: University of Chicago Press.

Jordan, H. (1883). Zur Biogeographie der nördlich gemäßigten und arktischen Länder. *Biologisches Centralblatt, 3*, 174–180, 207–217.

Kirby, W., & Spence, W. (1826). *Introduction to entomology*. London: Longman, Rees, Orme, Brown and Green.

Lacordaire, J. T. (1834). *Introduction à l'entomologie : comprenant les principes généraux de l'anatomie et de la physiologie des insectes, des détails sur leurs murs et un résumé des principaux systèmes de classification proposés jusqu'à ce jour pour ces animaux*. Paris: Pourrat frères.

Latreille, P. A. (1815). Introduction à la géographie générale des arachnides et des insectes; ou des climats propres à ces animaux. *Mémoire lu à l'Académie des Sciences, 3*, 37–67.

Linnaeus, C. (1737). *Flora Lapponicum, exhibens plantas per Lapponiuan crescentes*. London: James Edward Smith.

Marchant, J. (1916). *Alfred Russel Wallace: Letters and reminiscences* (2 vols.). London: Cassell.

Mayr, E. (1944). Wallace's line in the light of recent zoogeographic studies. *Quarterly Review of Biology, 19*, 1–14.
Mayr, E. (1946). History of the North American bird fauna. *Wilson Bulletin, 58*, 3–41.
Mennema, J. (1985). The first plant distribution map. *Taxon, 34*, 115–117.
Merriam, C. H. (1892). The geographical distribution of life in North America with special reference to the Mammalia. *Proceedings of the Biological Society of Washington, 7*, 1–64.
Merriam, C. H. (1899). *Results of a biological survey of Mount Shasta, California* (North American Fauna, 16). Washington, DC: United States Department of Agriculture, Division of Biological Survey.
Meyen, F. J. F. (1836). *Grundriss der Pflanzengeographie mit ausführlichen Untersuchungen über das Vaterland, den Anbau und den Nutzen der vorzüglichsten Culturpflanzen, welche den Wohlstand der Völker begründen*. Berlin: Haude und Spenersche Buchhandlung.
Meyen, F. J. F. (1846). *Outlines of the geography of plants*. London: Ray Society.
Minding, J. (1829). *Ueber die geographische Vertheilung der Säugethiere*. Berlin: Enslin'sche Buchhandlung.
Nelson, G. (1978). From Candolle to Croizat: Comments on the history of biogeography. *Journal of the History of Biology, 11*, 269–305.
Nicolson, M. (1987). Alexander von Humboldt, Humboldtian science and the origins of the study of vegetation. *History of Science, 25*, 167–194.
Nicolson, M. (1996). Humboldtian plant geography after Humbodlt: the link to ecology. *British Journal for History of Science, 29*, 289–310.
Nyhart, L. K. (2009). *Modern nature: The rise of the biological perspective in Germany*. Chicago: University of Chicago Press.
Ortmann, A. E. (1902). The geographical distribution of freshwater Decapods and its bearing upon ancient geography. *Proceedings of the American Philosophical Society, 41*, 267–400.
Ortmann, A. E. (1908). Bericht über die Fortschritte unserer Kenntnis von der Verbreitung der Tiere (1904–1907). *Geographisches Jahrbuch, 31*, 231–140.
Packard, A. S. (1883). *Monograph of Phyllopod crustacea*. Twelfth annual report of the U.S. Geological and Geographical Survey of the Territories, *1*, 295–592.
Parenti, L. R., & Ebach, M. C. (2009). *Comparative biogeography: Discovering and classifying biogeographical patterns of a dynamic earth*. Berkeley: University of California Press.
Prichard, J. C. (1826). *Researches into the physical history of mankind* (2nd ed.). London: Houlfton and Stoneman.
Ramond de Carbonnières, L.-F. (1798). Bemerkungen über die Vegetation auf den Gipfeln der höchsten Berge, besonders auf den südlichen Pic der Pyrenäen; wie auch über diejenigen Pflanzen welche mehrere Jahre lang sich unter dem Schnee erhalten können. *Neues polytechnisches Magazin. Eine uswahl aus den wichtigsten französischen Zeitschriften, 1*, 35–53.
Ritter, C. (1806). *Sechs Karten Von Europa: mit erklärendem Texte, darstellend: I. Die Verbreitung der Kulturgewächse in Europa. II. Die Verbreitung der wildwachsenden Bäume und Sträuche in Europa. III. Die Verbreitung der wilden und zahmen Säugethiere in Europa. IV. Die Hauptgebirgsketten in Europa, ihren Zusammenhang und ihre Vorgebirge. V. Die Gebirgshöhen in Europa, ihre Vegetationsgrenzen und verschiedenen Luftschichten; verglichen mit denen der heissen Zone. VI. Areal-Grösse, Volksmenge, Bevölkerung und Verbreitung der Volksstämme in Europa*. Schnepfenthal: Buchhandlung der Erziehungsanstalt.
Robinson, A., & Wallis, H. (1967). Humboldt's map of isothermal lines: a milestone in thematic cartography. *The Cartographic Journal, 4*, 119–123.
Schmarda, K. L. (1853). *Die geographische Verbreitung der thiere*. Vienna: Carl Gerold and Son.
Schmarda, K. L. (1866). Die Theirgeographie und ihre Aufgabe. *Geographisches Jahrbuch, 1*, 402–427.
Schmidt, K. P. (1955). Animal geography. In: E. B. Babcock, J. W. Durham, & G. S. Myers (Eds.), *A century of progress in the natural sciences 1853–1953*. Published in Celebration of the Centennial of the California Academy of Sciences (pp. 767–794). San Francisco.
Schmithüsen, J. (1985). Vor- un Frühgeschichte der Biogeographie. *Biogeographica, 20*, 1–166.
Schouw, J. F. (1823). *Grundzüge einer allgemeinen Pflanzengeographie*. Berlin: Reimer.

Sclater, P. L. (1858). On the general geographical distribution of the members of the class Aves. *Journal of the Proceedings of the Linnean Society: Zoology, 2*, 130–145.

Sclater, P. L., & Sclater, W. L. (1899). *The geography of mammals*. London: Kegan Paul, Trench, Trübner & Co.

Serje, M. (2004). The national imagination in New Granada. In R. Erickson, M. A. Font, & B. Schwartz (Eds.), *Alexander von Humboldt: From the Americas to the Cosmos* (pp. 83–97). New York: Bildner Center for Western Hemisphere Studies The Graduate Center, The City University of New York.

Sharpe, R. B. (1893). On the zoo-geographical areas of the world, illustrating the distribution of birds. *Natural Science, 3*, 100–108.

Stromeyer, F. (1800). *Commentatio inauguralis sistens historiae vegetabilium geographicae specimen*. Göttingen: Heinrich Dieterich.

Swainson, W. (1835). *A treatise on the geography and classification of animals*. London: Longman, Brown, Green, and Longmans.

Thiselton-Dyer, W. T. (1878). Lecture on plant-distribution as a field for geographical research. *Proceedings of the Royal Geographical Society of London, 22*, 412–445.

Unger, F. (1852). *Versuch einer geschichte der pflanzenwelt*. Vienna: Wilhelm Braumüller.

von Hofsten, N. G. E. (1916). Zur älteren Geschichte des Diskontinuitätsproblems in der Biogeographie. *Zoologische Annalen Zeitschrift für Geschichte der Zoologie, 7*, 197–353.

von Humboldt, A. (1816). XCIII. On the laws observed in the distribution of vegetable forms. *Philosophical Magazine Series 1, 47*, 446–453.

von Humboldt, A. (1817). Sur les lignes isothermes. *Annales de chimie et de physique, 5*, 102–111.

von Humboldt, A., & Bonpland, A. (1807). *Voyage de Humboldt et Bonpland* (Première partie. Physique Générale, et relation historique du voyage. Premier Volume, Contenant Essai sur la Géographie des plantes, accompagné d'un Tableau physique des régions équinoxiales, et servant d'introduction à l'Ouvrage). Paris: Chez Fr. Schœll.

von Zimmermann, E. A. W. (1783). *Kurze Erklärung der zoologischen Weltcharte*. Leipzig: Johann Beckmann.

Vuilleumier, F. (1988). Biogeography from 1888 to 1988: How much progress in 100 years? In R. Elzen, K.-L. Schuchmann, & K. Schmidt-Koenig (Eds.), *Current topics in avian biology: Proceedings of the international centennial meeting of the Deutsche Ornithologen-Gesellschaft* (pp. 19–23). Bonn: Deutsche Ornithologischen-Gesellschaft.

Wahlenberg, G. (1812). *Flora lapponica*. Berlin: Berolini.

Wahlenberg, G. (1813). *De vegetatione et climate in helvetia septentrionali inter flumina rhenum et arolam*. Turici Helvetorum: Impensis Orell, Fuessli et Socc.

Wallace, A. R. (1863). On the physical geography of the Malay Archipelago. *Journal of the Royal Geographical Society, 33*, 217–234.

Wallace, A. R. (1876a). *The geographical distribution of animals; with a study of the relations of living and extinct faunas as elucidating the past changes of the earth's surface*. London: Macmillan.

Wallace, A. R. (1876b). *Die geographische Verbreitung der Thiere* (Deutsche Ausgabe von A.B. Meyer). Dresden: Verlag R. von Zahn.

Wallace, A. R. (1883). On the value of the 'Neoarctic' [sic] as one of the primary zoological regions. *Nature, 27*, 482–483.

Wallace, A. R. (1894). What are zoological regions? *Nature, 49*, 610–613.

Watson, H. C. (1835). *Remarks on the geographical distribution of British plants*. London: Longman, Rees, Orme, Brown, Green, and Longman.

Watson, H. C. (1836). Observation on the construction of maps for illustrating the distributions of plants, with reference to the communication of Mr. Hind on the same subject. *The Magazine of Natural History, and Journal of Zoology, Botany, Mineralogy, Geology and Meteorology, 9*, 17–21.

Watson, H. C. (1847). *Cybele Britannica; Or British plants and their geographical relations* (Vol. 1). London: Longman & Co.

Watson, H. C. (1847–1859). *Cybele Britannica; or British plants and their geographical relations* (Vol. 4). London: Longman & Co.

Wilkins, J. S., & Ebach, M. C. (2014). *The nature of classification: Relationships and kinds in the natural sciences*. Basingstoke/New York: Palgrave Macmillan.

Willdenow, C. L. (1792). *Grundriss der Kräuterkunde*. Berlin: Haude and Spener.

Williams, D. M. (2007). Ernst Haeckel and Louis Agassiz: Trees that bite and their geographical dimension. In M. C. Ebach & R. Tangeny (Eds.), *Biogeography in a changing world* (pp. 1–59). Boca Raton: CRC Press.

Williams, D. M., & Ebach, M. C. (2008). *Foundations of systematics and biogeography*. New York: Springer.

Zimmermann, E. A. W. (1777). *Specimen zoologiae geographicae, Quadrupedum domicilia et migrationes sistens*. Leiden: Theodorum Haak.

Zimmermann, E. A. W. (1780). *Geographische geschichte des menschen, und der allgemein verbreiteten vierfüssigen thiere* (Vol. 2). Leipzig: Weygandschen Buchhandlung.

Zimmermann E. A. W. (1778–1783). *Geographische geschichte des menschen, und der allgemein verbreiteten vierfüssigen thiere* (Vol. 3). Leipzig: Weygandschen Buchhandlung.

# Chapter 6
# The Legacy of Nineteenth Century Plant and Animal Geography

By the beginning of the eighteenth century, Australia remained one of the few continents settled, by westerners, to remain unexplored. Of the many expeditions aimed at exploring the interior of Australia, Prussian expatriate and botanist, Ferdinand von Mueller was responsible for collecting, identifying and describing species. In his *Botanical Report on the North-Australian Expedition, under the command of A. C. Gregory, Esq.*, von Mueller described the regionalisation of Australia's plant life:

> It would lead beyond the limits of this document to contemplate the botany of the country in its full details, but I may sketch the principal distinctive features of the vegetation, which in a comprehensive view can be divided into the following groups:

1. Plants of the dense coast forests.
2. Plants of the Brigalow scrub.
3. Plants of the open downs.
4. Plants of the desert.
5. Plants of the sandstone table-land.
6. Plants of the sea-coast.
7. Plants of the banks and valleys of rivers (von Mueller 1858, p. 146, see also Ebach 2012).

By 1858 plant geography was in full swing in Europe and the Americas, with Schouw and Sclater having named their regions based on vegetation forms within larger geographical regions. The attempt by von Mueller, however, seems purely geographical, with no vegetation types, isotherms, or even rainfall mentioned. After all von Mueller did call it a sketch and never returned to it as a separate study. The same is true for English botanist Joseph Hooker, who lamented:

> ... [t]here are no geographical or other features of the Australian continent which enable me to draw any natural boundary between temperate and tropical Australia. In selecting a botanical tropic of Capricorn, I hence have had recourse to the distribution of the plants themselves, and these must afford very vague data (Hooker 1859, p. xxxviii).

Even attempts at drawing some boundary, like that of explorer and politician Sir George Grey were rejected by Hooker:

> The parallel of Sharks [sic] Bay, I have hence assumed to be north of the position of the tropic of vegetation [...] In determining what may be called the tropic of vegetation, regard must be had not only to the latitude and isothermal lines, but to the abundance of the vegetation and its character (Hooker 1859, p. xxxviii).

By the 1860s, Australian regionalisation was still in its infancy. Hooker instead referred readers to von Mueller's seven plant regions, something that the botanising Prussian himself had forgotten by 1882:

> The geographical limitations in this work coincide with the political boundaries of the colonial territories, except that the tropic of Capricorn eastward to the 188th degree separates what is here called Northern Australia (N.A.), from the South- and West-Australian extratropic possessions. Such geographic segregations are necessarily quite arbitrary, though they serve our present purpose of assigning to each of the colonial divisions of Australia its number of specified plants; the limitation is the same as that adopted in the Flora Australiensis, and as regards abbreviations also identical with the method of indications, chosen for the list of Australian trees in 1866 (von Mueller 1882, p. viii).

By 1882 there were still no Australian plant or animal regions that were universally acknowledged. The maps of Schouw, Berghaus and Grisebach were too broad, merely separating out Australia from New Zealand and New Guinea. The methods proposed by Schouw and Grisebach were also not used. Much of the Australian vegetation was unknown and the numbers of botanists able to identify and describe new species (and vegetation) were few and far between. Naturalist Ralph Tate drew the first phytogeographical regionalisation of Australia, who based his regions on a *Rain Map of Australia* (Fig. 6.1):

1. Euronotian (lit. south-east wind) dominant in the south and east parts of the Continent.
2. Autochthonian (lit. of the original race) restricted to the south-west corner of West Australia and approximately coinciding with the rain-fall limit of twenty inches.
3. Eremian (lit. desert) dominant the dry region, which has its centre in the Lake Eyre Basin [...] It is bounded on the north and north-east by the Indo-Australian vegetation; on the east and south-east by the typical Euronotian Flora, and on the extreme south-west by the Autochthonian (Tate 1889, p. 315).

Tate's regions were proposed in his Presidential address to the Australian and New Zealand Association for the Advancement of Science (ANZAAS) titled *On the influence of physiographic changes in the distribution of life in Australia*. Tate continued:

> I propose to make a beginning in the direction indicated by the foregoing citation, which of necessity concerns the geologist equally well as the botanist; believing, that however crude and imperfect our first efforts may be, they may nevertheless incite to further enquiry into all the circumstances involved and thereby advance to the attainment of our object more rapidly than if we permit the subject to be dormant until the said circumstances have been fully mastered independently (Tate 1889, p. 312).

Oscar Drude ignored the regions of Tate, von Mueller and Hooker, when he divided Australia into eleven regions:

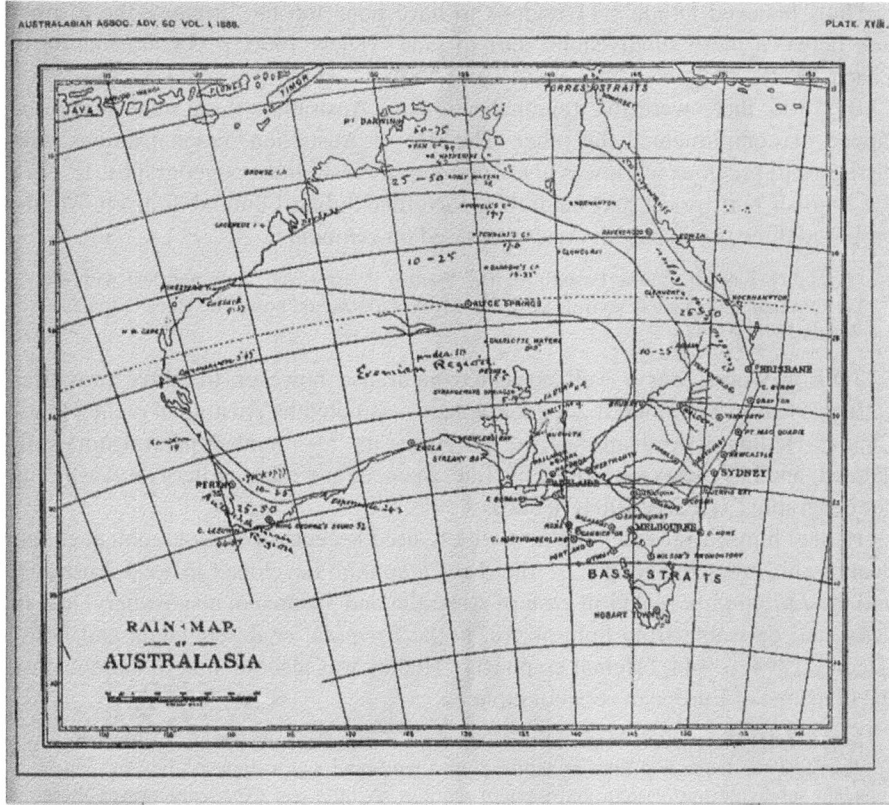

**Fig. 6.1** Ralph Tate's regions drawn onto the *Rain Map of Australasia* (Tate 1889, Plate XVIII) (Source: UNSW Australia Library)

North Australian region, Tropical Forest region, Queensland evergreen *Araucaria* and *Livistona* Forest region, North Australian Tree Savannah and Bushland region, Northwest Australian Transition region, West Australian Desert Steppe, East Australian Desert and Grass Steppe, Southwest Xerotide and Proteaceae region, South Australian Eucalyptus Forests, South Australian Eucalyptus and Fern region, the Mountain and Snow region of the Australian Alps, the Tasmanian Conifers, Grasslands and Mountain forest region (Drude 1890, pp. 499–502, translation in Ebach 2012).

Drude then, was perhaps the first plant geographer to apply the widely used vegetation system to divide up Australia's plants. Sixteen years later, fellow German Ludwig Diels in his *Die Pflanzenwelt von West-Australien* (Diels 1906). Diels divided Australia's flora into nine "Formations of Vegetation":

1. Tropical Rainforest, 2. Subtropical Rainforest, 3. Sclerophyll Forest, 4. Savanna Forest, 5. River Woodlands, 6. Beach Forests and Bushland, 7. Savanna, 8a. Mulga Scrub, 8b. Sub-littoral Sclerophyll Bushland, 8c. Heathland, 8d. Mallee Scrub, 8e. Brigalow Scrub, 9. Desert (Diels 1906, pp. 3–26, translation in Ebach 2012).

Diels believed Drude's 11 regions to have gone too far "because the distinction between these subdivisions start to fade" (Diels 1906, p. 38, translation in Ebach 2012).

By 1906, there were five regionalisations of Australia's flora, none that overlapped or complimented the other. The rise in Australian regionalisations with disregard to previous work was also common in Australian zoogeography. In 1878 the English born priest and naturalist, Reverend Julian Edmund Tenison-Woods, proposed three provinces for Australia based on echinoids:

> "1. The N. Eastern. 2. The Eastern. 3. The Southern". However, "I do not deal with the Western fauna, for I know so little of it, that my remarks would posses no value" (Tenison-Woods 1878, p. 147)

Tenison-Woods was a well respected naturalist, however, his three provinces were missed by a number of zoogeographers – as noted by Australian malacologist Charles Hedley, who dismissed the provinces as "... neither natural nor well-defined, and has been overlooked by Tate, Spencer and other writers on Australian zoogeography" (Hedley 1904, p. 880).

Hedley himself provided the first widely used accepted zoogeographical classification, his areas included "... the *Autochthonian*, developed in west Australia, and the *Euronotian*, seated in eastern Australia and Tasmania; a subsidiary, less in value and derivable from both above, is the *Eremian*, or desert fauna and flora" (Hedley 1894, p. 444, original emphasis). Hedley was also the first to comment on the divisions of European zoogeographers:

> [m]ost European writers who have touched on the zoo-geography of Australia have described the fauna and flora as falling into a temperate and a tropical division, which again subdivide into eastern and western sections. A little real experience proves these divisions to be quite artificial (Hedley 1893, p. 189, also reiterated in Hedley 1894, p. 444, see Ebach 2012).

Baldwin Spencer's revised Hedley regions in his 'Report on the work of the Horn Expedition to Central Australia' in 1896 (Fig. 6.2). In it, Spencer renamed and revised all of Hedley's regions:

> Torresian sub-region. This includes Papua and north and north-eastern Australia as far north as the Clarence River. On its north-western side it merges as might be expected to a certain extent into the western area [...] The Bassian sub-region. This includes the eastern and south-eastern coastal strip, lying between the coast line and the Dividing Range south of the Clarence River, and also Tasmania. On the mainland it naturally merges to a certain extent, where the dividing Ranges falls away at its western end, with the fauna of the interior but in the main it is strikingly dissimilar to this [...] The Eyrean sub-region. This includes the whole of the interior, southern and western part of the continent, the coastal region on the east and south- east separating it from the Torresian subregion in the north-east and the Bassian sub-region in the south- east (Spencer 1896, pp. 196–199).

Australian plant and animal geography is a good example of late nineteenth century practise. At the time there was no single or unified methodology. Practitioners from different backgrounds provided different area classifications. Diels, an ecologist divided up Australia's flora into vegetation types. Hooker and Hedley, both taxonomists, divided it along the distribution of taxa. Professionalism also

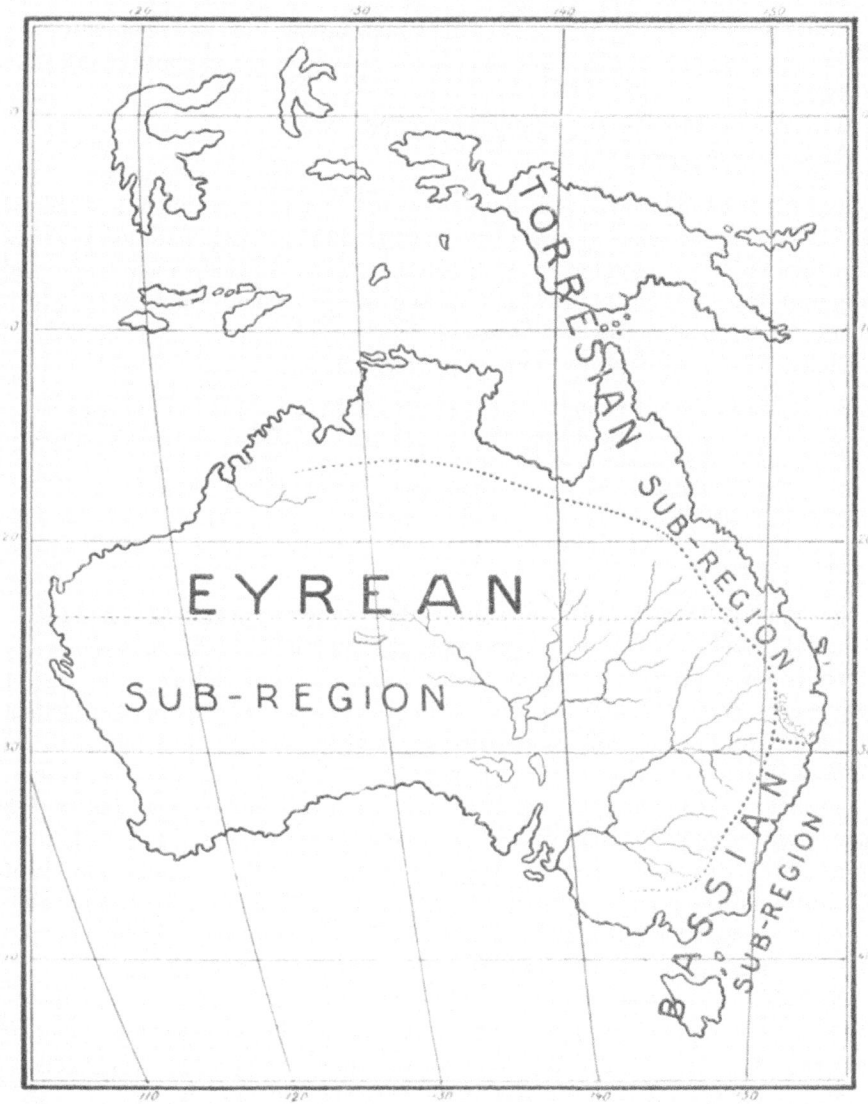

**Fig. 6.2** Walter Baldwin Spencer's *Faunal sub-regions of the Australian region* (1896) (Source: UNSW Australia Library)

had a role in regionalisation: Hedley's zoological regions, borrowed from Tate's phytogeographical regions, were renamed by Spencer to reflect zoogeographical regions. The legacy of nineteenth century plant and animal geography can be classified into five categories:

1. A misreading or avoidance of the literature
2. Professionalism driving both plant and animal classification and regionalisation
3. The division and conflict between taxonomic and vegetation classification systems
4. The resulting difference in regionalisation and;
5. The attempt at unification.

Together these five categories define how nineteenth century plant and animal geography was practised (i.e., organismal classification and regionalisation), rather than theorised (i.e., hypothesis of distribution, such as centres or origin and evolution). These five categories also have modern equivalents or analogies to which twenty-first century biogeographers can relate.

English geology Charles Lyell once declared that,

> [Buffon] the great French naturalist caught sight at once of a general law in the geographical distribution of organic beings, namely the limitation of groups of distinct species to regions separated from the rest of the globe by certain natural barriers. It was, therefore, in a truly philosophical spirit that, replying on the clearness of the evidence obtained respecting the larger quadrupeds, he ventured to call in question the identifications announced by some contemporary naturalists of species of animals said to be common to the southern extremities of America and Africa (Lyell 1842, p. 112).

Note that Lyell may be misread, namely "the great French naturalist caught sight at once of a general law in the geographical distribution of organic beings, namely the limitation of groups of distinct species to regions separated from the rest of the globe by certain natural barriers". Upon reading this passage it is clear that Buffon is referring to a *law of distribution*, namely that certain natural barriers control distribution that in turn lead to new species, rather than a taxonomic law. Modern scientists and historians refer to this as a distribution law one that Buffon "recognised for practical reasons each geographical region must possess different forms of life" (Browne 1983, p. 24). The view of Browne is the conventional view of most historians and scientists. But if we were to ignore the evolutionary implication of Buffon's statement and read further into the "practical reasons" for his law, and beyond the text given to us by Lyell, we find something else entirely.

The difference between the use of morphology (taxonomic groups) and climate and disposition (animal and plant forms) is a recurring theme in plant and animal geography. However, misreading or misinterpreting the literature also happens between contemporaries, such as Humboldt, Stromeyer and de Candolle.

The conflict between two practitioners of plant geography in the early nineteenth century, namely Humboldt and Stromeyer, is worthy of comment as it dragged in a third protagonist, A.P. de Candolle. The dispute was over priority. Humboldt had spent a third of his inheritance traveling the world, collecting in Latin America, between 1799 and 1804. During that time he climbed Mount Chimborazo, measured barometric pressure, temperature, rainfall, recorded astronomical observations, collected numerous plant specimens, sailed down the Amazon, and published what was at the time the most comprehensive study of Latin American geography, the 30 volume *Le voyage aux régions équinoxiales du Nouveau Continent, fait en*

*1799–1804* (Humboldt and Bonpland 1814–1829). In the *Essai*, Humboldt proposed a geography of plants, one that he had carefully devised during his travels. However, it wasn't until 1814, through a talk given by A.P. de Candolle at the *Arcueil* in Paris, that Humboldt was made aware of Stromeyer's *Specimen*, a revision and classification of plant geography. Humboldt's first reaction was to ask de Candolle to place it before his own *Essai* the 1814 manuscript of the talk. It seems quite certain that Humboldt had not read Stromeyer's *Specimen*, mainly because of his misspelling of Stromeyer's name and through his accusation that it did not contain any original measurements. Had Humboldt read the work, he would have noticed that it was a revision of plant geography, an attempt at classifying the field into its various parts based on what plant geographers studied. Unfortunately, it appears that Humboldt had not realised this until much later, possibly 1816 when he accurately evaluated Stromeyer's work. Why would someone held in a high esteem as Humboldt feel so threatened by a doctoral thesis?

The notion of priority is paramount to present day scientists. Priority translates into authority. In the early nineteenth century, plant geographers would hold some form of priority, as they were the authorities on their subject, meaning that they produced the next generation of doctoral students who would carry on the system of priority and authority. Take Stromeyer for example. While he proposed a large volume of work in his doctoral thesis, his *Specimen*, he did not follow through. Rather he went on to specialise in chemistry and in turn produce notable doctoral students such as Robert Bunsen. Humboldt was not in that system. He had no doctoral degree and was not an authority, but rather a polymath in an age of specialists. Stromeyer's thesis was essentially to make plant geography a specialist science, one in which Humboldt and his ideas did not feature greatly. Humboldt wanted priority and by 1823, when Schouw published his *Grundzüge*, had effectively cornered the field. Nonetheless, Humboldt's criticisms of Stromeyer and his badgering of de Candolle about his "Mirbel" does indicate that Humboldt had indeed misread or even avoided reading Stromeyer's *Specimen*.

Misreading the literature may be a result of presentism. Many modern professional ecologists, biogeographers and taxonomists seem to want to claim certain historical figures as founders of their field. For example, Gustaf Einar Du Rietz claims that:

> [Linnaeus] was a pioneer not only in taxonomy and morphology but also in genetics, dispersal ecology and phytogeography [and] was also one of the founders of Phytogeography (Du Rietz 1957, pp. 161–166).

Clearly du Rietz has a very broad definition of phytogeography. The way Linnaeus practised plant geography was very different to that of du Reitz or of any biogeographers in the 1950s. However, claiming such an important historical figure, regardless what they practised brings prestige as well as authority to a particular field. For example, Du Rietz was a Swedish ecologist (phytosociologist) who had penned such classics as *Factors Controlling the Distribution of Species in Vegetation* (Du Rietz 1929) and *Classification and Nomenclature of Vegetation* (Du Rietz 1930). Clearly du Rietz had adopted phytosociology (the study of vegetation types), rather than the Linnaean approach to plant classification, something that would have been abhorrent to Linnaeus. But yet, du Rietz claims Linnaeus as one of the "founders of Phytogeography", that is, of du Rietz's kind of phytogeography:

> ... the contributions of LINNAEUS to Phytogeography, this concept is taken in its widest sense, including *Phytosociology* (Du Rietz 1957, p. 161, original emphasis).

Specialisation also saw taxonomists using taxonomic distributions to define their regions, while physiologists spent their time looking for physiological forms to define their areas of vegetation. The result was a dual classification system: on one hand you had a strict adherence to a natural classification of taxa, while on the other you had a natural classification of vegetation forms. These both contradicted one another as they rarely overlapped and analogous forms were grouped together. The result in plant and later in animal geography was a distinct split in how geography was practised. Vegetation geographers were interested in climatic variables both in attitude and altitude. Taxonomic geographers were interested in actual species, genus and family distributions and avoided present day climate, as they believed modern distributions are results of past processes. This major conceptual split resulted in an interesting form of practise. Vegetation geographers worked on small-scale areas, while taxonomic geographers looked at large scale regions and realms. Perhaps A.P. de Candolle noticed this in 1820 when he proposed his stations and habitations. In any case, classifying two distinct practises in plant and animal geography as historical (large scale) and ecological (small scale) biogeography (*sensu* Nelson 1978) is both accurate but uninformative. The scale of the areas in question does not necessarily reflect the field of study. Vegetation geographers also look at large scale areas, which they define using existing geographical units, such as continents. The same is true for taxonomic geographers who may look at smaller highly endemic regions like Madagascar or southwestern Western Australia as sub-regions. The division between what Nelson (1978) described as historical and ecological biogeographers is clearly a division in how one classifies organisms and their associations with other organisms.

Classification is what drives plant and animal geography in the eighteenth, nineteenth and twentieth centuries and what still drives historical and ecological biogeography today (see Williams 2006; Kohler 2008). But that is no reason to use a presentist perspective on past practises. For example, when discussing one of the founders of ecological history, Cittadino noted:

> He [Donald Worster 1977] uses phrases like "eighteenth-century ecology" or "Linnaeus' ecology," which give the uninformed reader the mistaken impression that there was anything like a science of ecology in the eighteenth century (Cittadino 1979, p. 45).

In our attempt to communicate our ideas to our peers we revert to an unwitting form of presentism; that is, using modern terms with modern definitions to describe past practises. For example, Humboldt may have practised roughly what a twenty-first century ecologist would call "ecology", but the term itself, and the implied disciplinary independence, wasn't coined until 1861 by Ernst Haeckel. The modern definition or practise of ecology would not resemble Humboldt's.

In effect, Humboldt practised something other than ecology, which much later may be described as being "ecological".

Another example of a presentist reading of the literature is that of Miracle (2008). Miracle claims that nineteenth century Dutch zoologist "[Coenraad Jacob] Temminck formulated [...], for the first time, a general law concerning the geographical distribution of animals on the globe" (Miracle 2008, p. 681). The law states that,

> ... there is a relation in organisation, external form and behaviour between almost all animals that inhabit the latitude, however far from each other may be the regions where they live or freely disperse. The extension of the seas between regions has no influence in this, and neither has the enormous space of unbroken land that stretches between them (Temminck 1842, vol. 1, p. 7, in Miracle 2008, p. 685).

Miracle compares both Temminck and Buffon's laws as laws of distribution, only to find they are in fact "exactly the opposite" (Miracle 2008, p. 693). The contradictory views between Temminck and Buffon become apparent when we view them by their definitions, rather than as "distributional laws". Temminck's Law states how climate and geography has a role in shaping the same species no matter how widely dispersed. Buffon's Law, however, states how animal behaviour and resulting adaptions can be used to classify organisms in a non-hierarchical system of "genera". The Laws conflict because they are designed to do different things. Where then does Miracle get the idea that Buffon's and Temminck's Laws are the same? A footnote gives us a clue: "(Lyell in Nelson 1978, p. 274)" (Miracle 2008, p. 693, footnote 56). Miracle goes on to say that the "importance of Temminck's publications that described species distribution patterns relied not on their philosophical foundations [i.e., Temminck's Law], which definitely were not shared by other nineteenth century naturalists, but on their descriptive sections" (Miracle 2008, p. 701). This is correct. The practise of describing has far greater impact on the history of a field than its purported theories. If this is the case, then why do scientists and historians of science spend so much time discussing distributional theories when in fact they have no bearing on the practise of plant and animal geography?

I again return to specialisation in the natural sciences. Scientists rarely practise what they preach. A taxonomist may claim to be using a species concept, but in practise any notion of species concepts is far removed from the practise of describing a species. The same is true for distributional laws and area classification (i.e., regionalisation). Given much of eighteenth and nineteenth century plant and animal geography is the practise of classifying taxa or vegetation forms into areas, therefore any distributional hypotheses are moot. Laws belong to the general sciences, while the comparative sciences use classification (see Wilkins and Ebach 2014). However, by the late nineteenth century, distributional laws were mostly derived from the work of Darwin and Wallace (i.e., centre of origin, dispersal etc.). Regardless, the number of regionalisations had increased, rather than decreased, suggesting that there is a weak link between distributional laws and how we classify areas. Moreover, people like Wallace who championed the Sclaterian regions, supported a single origin for species, while Sclater supported the idea of multiple origins (polygenesis). Distributional laws, while interesting in their own right, had little to do with how plant and animals biogeographers had practised area classification. Rather it was the use of different organismal classification systems that had created a multiplicity of regionalisations.

## The Multidisciplinary Nature of Biogeography

Many claim biogeography as their own. For example, Frank Egerton in his review of Janet Browne's Secular Ark states that "Biogeography is a link between ecology and the earth sciences" (Egerton 1984, p. 405), thereby insinuating that biogeography is an ecological science (see also Egerton 2012, p. xiii). The claim is partially true. A history of biogeography from the 1980s would argue that biogeography is a link between biological systematics and earth history (Parenti and Ebach 2009). After all, much of taxonomic geography is based on the concept of natural taxa (i.e., natural classification). But attaining natural taxa and regions has been a problem, within the eighteenth, nineteenth and twentieth centuries as there was no method to find or test for natural classification. Taxonomists in denoting natural taxa have often resorted to authoritarianism, mostly by botanists and zoologists simply stating that their taxa are natural. Early ecologists like Grisebach (1872) noticed this absence of methodological rigour and instead moved toward highly quantified account of natural plant forms and vegetation. But proto-ecologists also suffered from a lack of unity in their method and in natural classification. By the early twentieth century they too started to resort to the same forms of authoritarianism, as did the taxonomists 100 years before. The two geographies that had emerged in the nineteenth century, lacked both methodological rigour and a single natural classification system. If practitioners from zoology, taxonomic botany and vegetation botany all introduce different approaches as to how areas are delimited and defined, we would find three different area classification systems, as we do in early Australian biogeography. However, if zoogeographers, for instance, propose single taxon classification systems, like that seen in Berghaus' maps between rodent and carnivore provinces, then we find there are more conflicting area classifications. Wallace's attempt to unify all zoogeographical regions into a single classification also backfired. A general classification of zoogeographical regions will be too broad scale for people working on single taxa in particular regions, like that of Allen and Merriam. For these American mammalogists, Sclaterian regions were poorly defined and the barriers too broad. Moreover, Wallace chose Sclaterian regions over Schmarda's regions, considering that Schmarda's regions were based on the same principles as those of Allen and Merriam, namely life zones. Given this, late 19th biogeography was headed towards multiplicity of classifications based on a multiplicity of ideas, approaches and organismal classifications. But can these approaches be classified into stations and habitations? If we divide both phyto- and zoogeography into the classification of organisms based on taxonomy and form, we find that whomever works on the classification of stations also works on the classification of habitations. Interestingly, those who propose classifications of habitations do not propose stations.[1] Given this, Nelson claimed that ecological and historical biogeography are split along the lines of stations and habitations. However, this was not the case for ecological biogeography, which consisted of both. By the early twentieth century this becomes more apparent. The German plant geographer Friedrich Ludwig Emil Diels was the first to perform a detailed study of Western Australian Flora and divided Western Australia into "nine Formations of vegetation" (see above). But Diels does not assign larger areas (i.e., habitations) and ignored the regions of Tate (1889). By 1906, Australian biogeography had split along the lines proposed by Nelson (1978). The ecologists were indeed working on stations, vegetation or "Formations of vegetation". In 1933, Professor George Edward Nicholls, a zoologist interested in historical biogeography, was the first to announce a regionalisation of both Australian flora and fauna in his Presidential Address at the ANZAAS Congress:

> In an attempt at generalization from such a set of facts as have been brought together in this survey it behoves on to go cautiously. There are numerous pitfalls in the path of the zoogeographer. For all but the specialist, in any given group, the actual identity or distinctness of named forms is frequently in doubt; in different orders, genera may come to have widely different values. The supposed facts of present day (or recent) distribution may prove to be insecurely based, locality records for material collected, for example, in the early days of Australian colonies being frequently vague, often quite erroneous (Nicholls 1933, p. 131).

Nicholls simply merged Hedley's and Spencer's regions with those of Tate, thereby creating a regionalisation that could be used by all biogeographers. But the split between ecologists and historical biogeographers remained, who both dutifully ignored each other's regions up until the 1990s (see Ebach 2012).

The multidisciplinary nature of biogeography stems from the biogeographers, the classification systems they use in zoology and botany, the methodologies employ and the theories they use. All these are transferred to biogeography creating what is effectively an eclectic field that is unifying in name only. Perhaps it is not surprisingly that Jordan, Merriam and Ratzel never bothered to define the term biogeography, because they too knew that biogeography (or animal and plant geography) did not have a common methodology and could only be unified by the predominant field of their day. Ratzel thought geography and geographical methods would unify biogeography. In the twentieth century the calls for unity were made along methodological lines (e.g., Donoghue and Moore 2003). In any case, any attempt at unity will only sideline a large majority of practitioners calling themselves "biogeographers". Unify biogeography under the auspices of geography and you isolate the biologists and palaeontologists. Unify biogeography under a methodology and you isolate those that use other methods and have differing aims.

---

[1] For example, de Candolle (1820), Prichard (1826), Sclater (1858), Wallace (1876), who would be considered historical biogeographers, did not propose any stations.

Today's problem of a multidisciplinary biogeography, with multiple aims, methods, theories and practises is not new and can be traced back as far as the beginning of the nineteenth century. As a multidisciplinary field, biogeography has multiple origins, hence the title of this book. The origins of late nineteenth century biogeography may be broken down further than proposed by Nelson (1978), to that of four fields.

1. Plant taxonomy and regionalisation (1750s–1820s) (Linnaeus, de Candolle)
2. Plant forms and vegetations (1780s–1920s) (Giraud Soulavie, Humboldt, Schouw, Meyen, Grisebach, Drude)
3. Animal taxonomy and regionalisation (1850s–Present day) (Linnaeus, Zimmermann, Prichard, Sclater and Wallace)
4. Animal forms and life zones (1850s–1920s) (Schmarda, Merriam, Allen).

Nineteenth century plant regionalisation I believed died with de Candolle (1820), as twentieth century plant biogeographers went on to use different regionalisations, namely that of Stanley Cain (1944) and Ronald Good (1947). The use of animal forms and plant forms clearly paved the way for early twentieth century ecology, which was slowly replaced by populations and population dynamics. The longest surviving practise is that of zoogeographical regionalisation, as Wallace's Sclaterian regions are still in use today (see Holt et al. 2013). What then about Australia's bioregionalisation?

The areas proposed by Tate, Hedley and Spencer, which were championed by Nicholls, were abandoned by the late 1990s for more quantitative approaches (see Ebach 2012). Ironically however, it was those quantitative approaches that have given these old areas a new lease of life. For example, in 2014, a geospatial analysis was conducted using "the largest digitized dataset of land plant distributions in Australia assembled to date (750,741 georeferenced herbarium records; 6,043 species) was used to partition the Australian continent into phytogeographical regions" (González-Orozco et al. 2014, p. 1). The analysis had uncovered the plant regions of Tate (1889) in spectacular detail. The regionalisations proposed over 100 years ago had paid off, only because the development of fast computer hardware and, intense collecting of data over the last 120 years had accumulated a data set large enough to be analysed confidentially. Nineteenth century regionalisation is still with us today.

## References

Browne, J. (1983). *The secular ark: Studies in the history of biogeography*. New Haven: Yale University Press.
Cain, S. A. (1944). *Foundations of plant geography*. New York: Harper and Brothers.
Cittadino, G. (1979). Nature's economy: The roots of ecology by Donald Worster. *Environmental Review, 3*, 1–4.
de Candolle, A. P. (1820). Essai élémentaire de géographie botanique. In *Dictionnaire des Sciences Naturelles* (Vol. 18, pp. 1–64). Paris: F. Levrault.

# References

Diels, L. (1906). Die Pflanzenwelt von West-Australien südlich des Wendekreises: mit einer Einleitung über die Pflanzenwelt Gesamt-Australiens in Grundzügen. In O. Drude & A. Engler (Eds.), *Vegetation der Erde VII.* Leipzig: W. Engelmann.

Donoghue, M. J., & Moore, B. R. (2003). Toward an integrative historical biogeography. *Integrative and Comparative Biology, 43*, 261–270.

Drude, O. (1890). *Handbuch der Pflanzengeographie.* Stuttgart: J. Engelhorn Verlag.

Du Rietz, G. E. (1929). The fundamental units of vegetation. *Proceedings of the International Congress of Plant Sciences, 1*, 623–627.

Du Rietz, G. E. (1930). The fundamental units of biological taxonomy. *Svensk Botanisk Tidskrift, 24*, 333–428.

Du Rietz, G. E. (1957). Linnaeus as a phytogeographer. *Vegetatio, 5*, 161–168.

Ebach, M. C. (2012). A history of biogeographical regionalisation in Australia. *Zootaxa, 3392*, 1–34.

Egerton, F. N. (1984). The secular ark: Studies in the history of biogeography by Janet Browne. *Isis, 75*, 405–406.

Egerton, F. N. (2012). *Roots of ecology: Antiquity to Haeckel.* Berkeley: University of California Press.

González-Orozco, C. E., Ebach, M. C., Laffan, S. W., Thornhill, A. H., Knerr, N. J., Schmidt-Lebuhn, A. N., Cargill, C. C., Clements, M., Nagalingum, N. S., Mishler, B. D., & Miller, J. T. (2014). Quantifying phytogeographical regions of Australia using geospatial turnover in species composition. *PLoS ONE, 9*, e9258.

Good, R. (1947). *The geography of flowering plants.* New York: Longmans, Green and Co.

Grisebach, A. H. R. (1872). *Die Vegetation der Erde nach ihrer klimatischen Anordnung.* Leipzig: Wilhelm Engelmann.

Hedley, C. (1893). On the relation of the fauna and flora of Australia to those of New Zealand. *Natural Science, 3*, 187–191.

Hedley, C. (1894). The faunal regions of Australia. *Reports of the Australian Association for the Advancement of Science, 5*, 444–446.

Hedley, C. (1904). The effect of the Bassian isthmus upon the existing marine fauna: A study in ancient geography. *Proceedings of the Linnean Society of New South Wales, 28*, 876–883.

Holt, B. G., Lessard, J. P., Borregaard, M. K., Fritz, S. A., Araújo, M. B., Dimitrov, D., Fabre, P. H., Graham, C. H., Graves, G. R., Jønsson, K. A., Nogués-Bravo, D., Wang, Z., Whittaker, R. J., Fjeldså, R. J., & Rahbek, C. (2013). An update of Wallace's zoogeographic regions of the world. *Science, 339*, 74–78.

Hooker, J. D. (1859). *The flora of Australia, its [sic] origin, affinities, and distribution; being an introductory essay to the Flora of Tasmania.* London: Lovell Reeve.

Kohler, R. E. (2008). Plants and pigeonholes: Classification as a practise in American ecology. *Historical Studies in the Natural Sciences, 38*, 77–108.

Lyell, C. (1842). *Principles of geology or the modern changes of the earth and its inhabitants, considered as illustrative of geology by Charles Lyell* (Vol. 3). Boston: Hilliard, Gray & Co.

Miracle, M. E. G. (2008). The significance of Temminck's work on biogeography: Early nineteenth century natural history in Leiden, the Netherlands. *Journal of the History of Biology, 41*, 677–716.

Nelson, G. (1978). From Candolle to Croizat: Comments on the history of biogeography. *Journal of the History of Biology, 11*, 269–305.

Nicholls, G. E. (1933). The composition and biogeographical relations of the fauna of Western Australia. *Reports of the Australian Association for the Advancement of Science, 21*, 93–138.

Parenti, L. R., & Ebach, M. C. (2009). *Comparative biogeography: Discovering and classifying biogeographical patterns of a dynamic Earth.* Berkeley: University of California Press.

Prichard, J. C. (1826). *Researches into the physical history of mankind* (2nd ed.). London: Houlfton and Stoneman.

Sclater, P. L. (1858). On the general geographical distribution of the members of the class Aves. *Journal of the Proceedings of the Linnean Society: Zoology, 2*, 130–145.

Spencer, W. B. (1896). *Report on the work of the Horn Scientific Expedition to Central Australia: Part 1 – Introduction, narrative, summary of results, supplement to zoological report, map.* Melbourne: Melville, Mullen & Slade.

Tate, R. (1889). *On the influence of physiological changes in the distribution of life in Australia*, Report of the first meeting of the Australian Association for the Advancement of Science (pp. 312–326).

Temminck, C. J. (1842). *Coup-d'oeil général sur les possessions néerlandaises dans l'Inde archipélagique.* Leiden: Arnz.

Tenison-Woods, J. E. (1878). The echini of Australia. *The Proceedings of the Linnean Society of New South Wales, 2*, 145–176.

von Humboldt, A., & Bonpland, A. (1814–1829). *Le voyage aux régions équinoxiales du Nouveau Continent, fait en 1799, 1800, 1801, 1802, 1803 et 1804.* Paris: Librairie grecque-latine-allemande.

von Mueller, F. (1858). Botanical report on the North-Australian expedition, under the command of A. C. Gregory, Esq. *Journal of the Proceedings of the Linnean, 2*, 137–163.

von Mueller, F. (1882). *Systematic census of Australian plants.* Melbourne: McCarron, Bird & Co.

Wallace, A. R. (1876). *The geographical distribution of animals; with a study of the relations of living and extinct faunas as elucidating the past changes of the earth's surface.* London: Macmillan.

Wilkins, J. S., & Ebach, M. C. (2014). *The nature of classification: Relationships and kinds in the natural sciences.* Basingstoke/New York: Palgrave Macmillan.

Williams, D. M. (2006 [2007]). Ernst Haeckel and Louis Agassiz: Trees that bite and their geographical dimension. In M. C. Ebach & R. Tangey (Eds.), *Biogeography in a changing world* (pp. 1–59). Boca Raton: CRC Press.

# Epilogue

Most books that cover some aspect of the history of biogeography focus on the history of ideas or is practitioners and rarely on the aims and methods, and how it relates to a twenty-first century biogeographer. I hope that my book has helped you, a twenty-first century biogeographer, understand the multidisciplinary nature of the field, why it can never be unified and how it has multiple origins, most of which had not survived the new methodological approaches of the early twentieth century. What is with us today that can truly be said to be "19th century" is regionalisation; carving the world into natural regions, whatever they may be.

As for the historian of science, I have not engaged in your literature, nor have I used any historical methods. This work is a *science* historiography, one that focuses on the works of *scientists* rather than historians. I am a biogeographer wanting to tell biogeographers a history of biogeographical practise. I hope that this will help biogeographers understand that their current situation is not new nor is it dire. Biogeography is not in disarray or in conflict. Biogeography is multidisciplinary, something which we are best to accept. Recent calls for unity are moot and appealing to historical founders or patriarchs unnecessary, given they were practicing a very different science.

The chapters in this book serve as vignettes of a much larger history. I have not included all practitioners of plant and animal geography to do so would result in much overlap. I have focused on a few good examples to explain a problem, method or issue. I have concentrated on specialisation in terms of how it affected the practise of plant and animal geography, rather than how it drove institutions. Famous figures like Darwin for example, have only a small role to play in the practise of plant and animal geography. Darwin was someone who used geography to advance an idea of evolution, rather than to contribute to nineteenth century plant or animal geography. In fact how organisms were classified and the role the environment played in shaping plant and animal forms, had a greater impact on nineteenth century plant and animal geography than did evolution or early ideas of continental drift. The biogeography we practise today is vastly different from that of the eighteenth and nineteenth century. Not many biogeographers understand this and often claim that Humboldt was an ecologist or Darwin was a biogeographer. I hope this book clarifies a lot of these misconceptions and helps the twenty-first century biogeography understand his eighteenth and nineteenth century colleagues and how they did plant and animal geography.

# Biosketches

The following biosketches summarise the contribution of several key historical figures in plant and animal geography as outlined in the above chapters. The following biosketches, arranged in alphabetical order, may be used to as a quick guide to the arguments outlined in the book. Names in the text highlighted in bold have are also listed as separate entries within the biosketches.

**Joel Asaph Allen (1838–1921)**
American ornithologist who studied under Louis Agassiz (1807–1873) and later became curator of birds and mammals at the American Museum of Natural History. Allen was opposed to the use of the six Sclaterian regions, as they were "based on continental areas, regardless of the actual distribution of life" and that the regions of **Phillip Lutley Sclater (1863–1913)** were "in opposition to nearly all other systems, whether of botanists or zoologists, who in general recognize that the distribution of life is in accordance with the climatic zones, in virtue of climatic influences, which the Sclaterian school consider as superficial and misleading" (Allen 1892, p. 212).

Allen along with fellow American mammalogist **Clinton Hart Merriam (1855–1942)** opposed the use of the Sclaterian system, as it was not based on climatic zones and therefore not a natural division of the earth's zoogeographical regions.

**Heinrich Karl Wilhelm Berghaus (1797–1884)**
Berghaus was a Prussian geographer who produced the *Physikalischer Atlas* between 1838 and 1848. The Atlas contains some of the very first zoogeographical regionalisations of vertebrates as well as reproductions of existing plant regions. The maps were unique as they separated the concept of distribution ("Verbreitung") from division ("Verteilung"), thereby showing animal and plant distributions as well as their regions. The maps of Berghaus are mostly translated from the work of **Joakim Frederik Schouw (1789–1852)** and **Eberhard August Wilhelm von Zimmermann (1743–1815)**. The *Physikalischer Atlas* of Berghaus was the most accessible regionalisations during the latter part of the nineteenth century.

### Georges-Louis Leclerc (Comte de) Buffon (1707–1788)
French naturalist, Buffon produced the monumental *L'Histoire Naturelle, générale et particulière, avec la description du Cabinet du Roi* (1749–1788). Buffon criticised the taxonomic system of **Carl Linnaeus (1707–1778)**, as disregarding the environments in which species are found. Rather, Buffon proposed a new system in which species are described based on their climate and disposition as opposed from their morphology alone, without the need for a hierarchical classification taxonomy. Buffon is famous for his so-called law of distribution, namely that similar environments separated over great distances do not produce the same species. Rather, Buffon's Law is a taxonomic law, one that is used to distinguish the differences between species from different areas for the purposes of identification and classification.

### Augustin Pyramus de Candolle (1778–1841)
Swiss botanist, A.P. de Candolle, was famous for his work on natural classification, particularly his 1813 *Théorie élémentaire de la botanique*, in which he formally proposed that plant morphology, rather than physiology, is the basis for a natural classification. In 1820 de Candolle published his *Géographie botanique* in the *Dictionnaire des Sciences Naturelles*. In it, he proposed the first plant regions of the world based on taxonomic distribution, the concept of plant endemism as well as dividing plant geography into two fields of study, those based on stations (habitats) and those based on habitations (regions). In 1805 in the *Flore française* (3rd edition), de Candolle published the first biogeographical map, titled *Carte Botanique de France,* namely a classification of the plant regions of France.

### Alphonse de Candolle (1806–1893)
The son of **Augustin Pyramus de Candolle (1778–1841)**, Alphonse Pyramus de Candolle is remembered for his 1855 *Géographie botanique raisonée* in which he rejected his father's system of regionalisation as descriptive and artificial.

### Carl Georg Oscar Drude (1852–1933)
German botanist and plant geographer, Oscar Drude proposed an autonomous biological system in his 1890 *Handbuch der Pflanzengeographie*. Drude's Natural vegetation classification system is a separate classification system of vegetative forms rather than on morphology as in the Linnaean system, which he considered artificial. Drude was influenced by **August Heinrich Rudolf Grisebach (1814–1879)** to whom he dedicates his 1890 "Handbuch".

### Heinrich Gustav Adolf Engler (1844–1930)
German botanist and plant geographer, Adolf Engler proposed a third system for plant geography that attempted to combine taxonomic plant geography and vegetation geography.

### Edward Forbes (1815–1854)
In 1854, English naturalist Edward Forbes proposed 25 zoogeographical provinces and nine Homoiozoic belts based on isothermal lines (along latitudinal gradients) and five bathymetric zones, in order to delimit the marine zoogeographical regions

of the world. While popular with naturalist like Joseph Hooker, Forbes' idea of dividing regions by Homoiozoic belts was not generally adopted in animal geography.

**Jean-Louis Giraud Soulavie (1751–1813)**
Giraud Soulavie was a French abbott (Abbé) and naturalist, published the first phytogeographical profile titled *Vertical cross-section of the Vivaroises Mountains; respective limits of plants* in his the eight volume *Histoire naturelle de la France méridionale* in 1783. The profile was unique as it identified plant regions of agricultural importance, such as the "Climate of Chestnuts" and the "Climate of Vines". A similar approach to drawing phytogeographical profiles was adopted by **Friedrich Wilhelm Heinrich Alexander von Humboldt (1769–1859)** in his 1805 *Essai sur la géographie des plantes*.

**August Heinrich Rudolf Grisebach (1814–1879)**
German botanist and plant geography, Grisebach championed the classification of vegetation forms (physiognomic groups) in order to establish natural regions of vegetation. In his 1872 *Die Vegetation der Erde nach Ihrer Klimatischen Anordnung*, Grisebach proposed "Geobotanik", namely the study of the vegetation of single countries, which influenced fellow German botanist **Carl Georg Oscar Drude (1852–1933)** and a generation of early plant ecologists.

**Friedrich Wilhelm Heinrich Alexander von Humboldt (1769–1859)**
Prussian explorer, botanist and aristocrat, Alexander von Humboldt spent two-thirds of his inheritance to fund his research including an expedition to the Americas (1799–1804). Humboldt's plant geography was a break away from describing species distributions, to one of proposing regions of vegetation, based on the physiognomic characteristics of plants. The famous *Tableau* of Humboldt, published in his 1805 [1807] *Essai sur la géographie des plantes*, is most likely adopted from the phytogeographical profiles of **Jean-Louis Giraud Soulavie (1751–1813)** *Vertical cross-section of the Vivaroises Mountains; respective limits of plants* of 1783 and Francisco José de Caldas' unpublished *Memoir on the distribution of plants that are cultivated near the equator*. While several phytogeographical profiles were published during the nineteenth century, Humboldt's enduring legacy is that of his classification of vegetation forms, which was refined by nineteenth century Humboldtians **Joakim Frederik Schouw (1789–1852)**, **Franz Meyen (1830–1840)** and **August Heinrich Rudolf Grisebach (1814–1879)**. Humboldt's *Kosmos*, published between 1845 and 1862 stand as a synthesis of his entire lives work.

**Hermann Jordan (18??–19??)**[1]
German animal physiologist (specialising in molluscs) and a *privatdozent* (adjunct lecturer). Jordan studied under Carl Eduard von Martens (1831–1904) and coined

---

[1] The exact dates of Jordan are unknown. Even the work *2,400 Years of Malacology* (10th edition) by Coan et al. (2013) states "Jordan, Hermann (18\*\*–\*\*\*\*; Germany). Fresh-water of Germany and Asia (1880s)" (2013, p. 491). I have been unable to find any reference to Jordan's life.

the term "biogeographie" in the German language in his 1883 work *Zur Biogeographie der nördlich gemäßigten und arktischen Länder*.

### Johann Heinrich Friedrich Link (1767–1851)
German botanist, director of the Berlin Botanical Garden and successor of Willdenow, Link completed his doctoral thesis *Flora der Felsgesteine rund um Göttingen* in 1789 in Göttingen under Johann Friedrich Blumenbach (1752–1840).

### Christian Menzel (1622–1701)
Also known as Christiani Mencelii (see Lesser 1751, p. 321), "... is said to have travelled a good deal on purpose to examine the different plants of his native country. He possessed likewise great skill in a variety of foreign languages, and was even well acquainted with the Chinese. Menzel was physician to his Majesty [Friedrich Wilhelm I, King of Prussia] at Berlin" (Willdenow 1811, pp. 464–465, however gives an incorrect date of 1662–1710). The German 1810 edition of Willdenow (Willdenow 1810, p. 548), however, gives Menzel's dates as 1622–1701 and is herein considered correct.

### Clinton Hart Merriam (1855–1942)
American zoologist Clinton Hart Merriam coined the term "biogeography" in the English language to describe "principal bio-geographic divisions" of the earth. In his 1892 *Geographical distribution of life in North America with special reference to mammals*, Merriam discussed the use of life-zone as opposed to the Sclaterian regions, which he believed were inaccurate and aimed to provide a convent classification. Merriam's own regions were based on the relative "numbers of distinctive types of mammals, birds, reptiles, and plants they contain" as well as the works of other practitioners like **August Heinrich Rudolf Grisebach (1814–1879)** and **Carl Georg Oscar Drude (1852–1933)**.

### Franz Julius Ferdinand Meyen (1830–1840)
Meyen was a Prussian plant physiologist and anatomist who adopted the Humboldtian vegetation classification. Meyen's 1836 *Grundriss der Pflanzengeographie* investigated how climate (i.e., weather) affects the distribution of plants. In it Meyen created a plant geography based on a non-hierarchical classification of floras, vegetation and forms that were driven by latitude and in altitude and delineated by isothermal lines.

### Alfred Newton (1829–1907)
English zoologist, ornithologist and correspondent of **Alfred Russel Wallace (1823–1913)**, Alfred Newton opposed the idea of a Nearctic and Palearctic, instead proposing a unified Holarctic.

### James Cowles Prichard (1786–1848)
Prichard, an English zoologists and ethnologist, proposed the first zoogeographical regions based on mammal distribution in his 1826 *Researches into the Physical History of Mankind*. The same work also proposed all human races are part of the same species, indicating a place of origin for all humans.

## Joakim Frederik Schouw (1789–1852)

Danish botanist, Schouw was the first plant geographer to adopt Humboldt's geography of plants. In his 1823 *Grundzüge einer allgemeinen Pflanzengeographie* (originally published in Danish as *Grundtræk til en almindelig Plantegeographie* in 1822), Schouw proposed plants regions based on Humboldtian principles, such as climate and the dominant vegetation form. Schouw work is the first application of Humboldt vegetation system in plant regionalisation.

## Ludwig Karl Schmarda (1819–1908)

Schmarda was an Austrian zoologist who proposed the first general zoogeographical regions of the world in 1853 in his *Die Geographische Verbreitung der Thiere*. Schmarda's regionalisation consisted of two sets of areas; one set was based on animal forms while the other mapped taxic distributions, like marsupials. What makes Schmarda's regions so unique is that they include land and marine regions. Schmarda also promoted the idea of animal forms based on the concept of variant forms, namely unrelated taxa that share the same physiology based on the environments in which they live.

## Philip Lutley Sclater (1863–1913)

Sclater, an English zoologist and ornithologist, is responsible for identifying the main zoogeographical realms and regions of the world. His 1858 paper *On the general geographical distribution of the members of the class Aves*, proposed six new zoogeographical regions and two realms, Palaeogeana (the Old World) and Neogeana (the New World). **Alfred Russel Wallace (1823–1913)** adopted Sclater's bird regions and revised them as vertebrate zoogeographical regions. Sclater's regions have been in constant use since their appearance in the mid nineteenth century and are used by twenty-first century biogeographers.

## Friedrich Stromeyer (1776–1835)

Stromeyer spent most of his early and latter years in Göttingen, then part of the Duchy of Brunswick-Lüneburg, where he obtained his doctorate degree in medicine under naturalist Johann Friedrich Gmelin. After travelling through France, Switzerland and studying under chemist Louis Nicolas Vauquelin in Paris, Stromeyer joined the faculty at the University of Göttingen in 1802, replacing his mentor who later died in 1804. At Göttingen Stromeyer pursued chemistry where he discovered cadmium in 1817 and mentored a new generation of German speaking chemists, most notably Robert Wilhelm Eberhard Bunsen (1811–1899). With such an influence and productive life in chemistry, it is no wonder that Stromeyer had abandoned plant geography, particularly given the reaction of practitioners like **Friedrich Wilhelm Heinrich Alexander von Humboldt (1769–1859)**.

## William John Swainson (1789–1855)

Swainson, and English zoologist, proposed natural zoogeographical regions based on the work of **James Cowles Prichard (1786–1848)**. In 1835 Swainson published his famous *A Treatise on the Geography and Classification of Animals*, in which he revised Prichard's regions. Swainson is notable in referring to natural regions based on the distribution of taxa. By 1835 many plant geographers considered regions based on taxonomic distributions to be arbitrary, conforming to current geographical areas rather than to "natural" forms characteristic of the environments they inhabit.

### Göran Wahlenberg (1780–1851)
Wahlenberg was a Swedish naturalist who in his 1812 *Flora Lapponica* proposed a regionalisation of plants based on taxic distributions. In 1813, Wahlenberg included a phytogeographical profile in his *De vegetatione et climate in helvetia septentrionali inter flumina rhenum et arolam*.

### Alfred Russel Wallace (1823–1913)
Celebrated English naturalist, Alfred Russel Wallace was responsible for the revival of the 1853 ornithological regions of Philip Lutley Sclater (1863–1913). **Clinton Hart Merriam (1855–1942)** and **Joel Asaph Allen (1838–1921)** challenged Wallace's regions, particularly the Nearctic and Palaearctic, which were considered to be the same region based on overlapping board and mammal distributions. Wallace's regions are still in use today and recent geospatial studies have recovered these areas, indicating that these are indeed naturally occurring zoogeographical regions.

### Carl Ludwig Willdenow (1765–1812)
Willdenow is regarded as one of the most important Prussian botanists of the eighteenth century. His 1792 *Grundriss der Kräuterkunde* was considered to be the bible of botany and much of his work on plant distribution influenced both **Friedrich Wilhelm Heinrich Alexander von Humboldt (1769–1859)** and **Friedrich Stromeyer (1776–1835)**. Unlike Humboldt and Stromeyer, Willdenow did not classify taxa into the regions they occur or, propose a phytogeographical method. Rather Willdenow simply referred to where plants were distributed and offered a theory of their distribution, in a similar way to Carl Linnaeus (1707–1778).

### Eberhard August Wilhelm von Zimmermann (1743–1815)
Prussian zoologist, mathematician and geographer, Eberhard August Wilhelm von Zimmermann, is perhaps the first person to classify zoogeographical regions based on animal distributions. In 1777 Zimmermann also produced the distribution map of quadrupeds (mostly mammals) *Tabula mundi geographico zoologica sistens quadrupedes hucusque notos sedibus suis adscriptos* in his *Specimen Zoologiae Geographicae Quadrupedum*. In doing so, Zimmermann is the first zoologist to propose distributional limits of mammals based on geographical phenomena, such as mountain chains, oceans and temperature.

# Appendix. Translation of the Introduction to "Commentatio Inauguralis Sistens Historiae Vegetablium Geographiae Specimen" by Friedrich Stromeyer (1800) (Translation by Mark Garland)[1]

Physical Society of Göttingen and Member of the Medical Society of Paris.

*"Plants are not randomly dispersed on Earth"* Jacques-Henri Bernardin de Saint-Pierre [1797, p. 289][2]

## Introduction

I. *What is the Geographic History of Vegetables, and what things and ideas are comprehended under this general denomination.*

The discussion of this matter rests upon these questions:

1. How the incredible multitude of Vegetables and their forms and conformations, endlessly various and multiform, spread and are distributed over the earth at the present day; what are the laws of their spread and distribution.-Geography of Vegetables, Phyto-geography.

---

[1] Stromeyer's introduction to his *Specimen* provides a detailed 80 page overview of the travel, work and writings of plant geographers and taxonomists of the eighteenth century. The *Specimen* has a four page preface in which Stromeyer acknowledges Johann Friedrich Blumenbach and Georg Christoph Lichtenberg, followed by a nine page contents section and an introduction. The rest of the text is divided into two books, the first on current global plant distributions and, the second on former global plant distributions. The introduction (pp. 14–33) is translated into English for the first time below. Original pages number given in brackets. Original references given in footnotes have been placed into the references section and the year added to the relevant author in brackets.

[2] The original states "Les plantes ne sont donc pas jetées au hasard sur la terre, et quoiqu'on n'ait encore rien dit sur leur ordonnance en général dans les divers climats, cette simple esquisse suffit pour faire voir qu'il y a de l'ordre dans leur ensemble" Saint-Pierre (1797, p. 289, reprinted from *Études de Nature*).

2. Whether Vegetables formerly occupied the earth in the same way as at the present day, or if that is truly the case, and what they may have undergone before they arrived at that station in which they are now found; what causes provide a place for these changes, and what things follow from this. – Geographic History of Vegetables.
3. Lastly, by what relation all these things agree with the history of the earth and the rest of its inhabitants, both man and animals. – Applied Geographic History of Vegetables.

[p. 15] After the example of the celebrated Zimmermann [1777, 1778–1783], who so excellently published on a similar subject in the matter of animals, I believe that I can gather the whole outline of this subject, so full and abundant, scarcely inconveniently under this common notion, and encompass it by the denomination of the Geographic History of Vegetables.

II. *On those things which have been written on the Geographic History of Vegetables thus far, and on their sources.*

There is practically no other part of Botany treated with less zeal and more neglected and less esteemed than the Geographic History of Vegetables, and from that fact in no part of this science is our knowledge found to be so defective and imperfect as in this; inasmuch as up to now it has arrived at a very poor stage of elaboration. The anatomy and physiology of plants, though from the evil character of the time little cultivated and nearly always little cared for, have had nevertheless certain cultivators and promoters here and there. This truly has scarcely happened to the Geographic History of Vegetables, though it is a subject most worthy to be known and understood. [p. 16] For if we survey and fly through the long series of botanical writings, from the origin of the science up to the times of the great Hedwig and his contemporaries, we will light upon not one name which in any way can rival or follow Zimmermann.

A work of the same genius and classic worth is exactly desired, such as that great investigator of nature has composed and completed on the History of Man and the Geography of Quadrupeds; in which scattered arguments and evidence, whatever is known and observed about the geographic state of the Vegetables of our world and their history, are found in a single and entire work, with critical judgment, collected and compiled in order and with a philosophical view.

Only lately have there indeed appeared certain specimens of the Geographic History of Plants, which nevertheless in both the variety of subjects and the richness of inferences and the plan of treatment and exposition, in no way may be compared with Zimmermann's work. Moreover, even if judgment be brought sincerely and without any envy about these, it must be admitted that they are as far away as possible from satisfying expectations that one may justly and rightly conceive and hold about the treatment of such an excellent and ample matter. But truly I do not at all want it to be said that I seem to censure those essays for this reason, or to carp at and diminish their merits; but I only wanted to nod at how little they have esteemed the argument, which deserves a distinguished [p. 17] place in Botany, worthy by attention and study, than they themselves have given to the work for a short time. No one will deny that authors of those specimens deserve great praise, which were first, who cleared the way to this matter, and opened access to this spring to natural science, full of new explanations and notions, when moreover their writings attentively and diligently selected both cause delight and bring utility.

Sources are scarcely lacking from another part for working out the Geographic History of Vegetables. Not only do we possess some excellent commentaries concerning certain individual matters pertaining to it, beyond large catalogues of plants from many parts of the earth, but moreover from the leaders of the investigators of nature, as they happened into diverse regions, observations and proofs not at all unworthy of themselves have contributed to the Geography of Vegetables.

Meanwhile it is easily perceived that these sources are not equal in value: for both the largest catalogues of plants and very many observations of travellers from which we must especially [p. 18] draw, very often are somewhat imperfect, and not well defined and less accurately interpreted. For apart from the fact that in each one who instituted and treated observations about such a matter, neither the famous genius of Linnaeus, Forskål, Pallas, Saussure, Ramond, or the Forsters, nor their great perspicacity, nor their ample learning are found united in happy marriage: also, very few disquisitions and observations have been intentionally made and annotated by their authors, so that many things might be brought and introduced for augmenting the Geographic History of Vegetables. Hence it has turned out that they have brought forward nothing but defective fragments. It is therefore necessary that, since we rarely hesitate in approaching and drawing from these sources, and observations of others may not complete the failings of these, either those things that are lacking from this subject may be completed by conjectures, or they may be left untouched.

Another influence, which equally has consideration in using these sources, resides in the fact that observations occur scattered and dispersed in the most numerous, diverse and various writings and works, and moreover are treated in diverse manner and language. He who will ever give any study to such things will not be ignorant of how much difficulty and labour will be brought to these things and how much work will be demanded, lest anything be overlooked that may be useful. It may be demonstrated that all these things have contributed much, at least in some part, even though so little has up to now been done in treating the Geographic History of Vegetables.

[p. 19] To this is finally added that, however huge may be the multitude of sources regarding the Geographic History of Vegetables, those regarding single parts are very poor, and observations are so distributed according to a most dissimilar plan, that we can scarcely hope that clear and evident notions will be exhibited to us; this holds, for example, in the migrations of Vegetables. And if we are not plainly to dismiss this argument, so weighty and of such moment, no other way lies open than to select and propose those things that approach the stage of greatest probability from little fragments, notices, and a series of conjectures, which a disquisition on the present extension and distribution of Vegetables supplies. However this may be, much light and knowledge will be able to be brought and given to this obscure part of the natural History of Vegetables, if all these things are rightly and cautiously made use of and diligently investigated.

It may hardly be without advantage and not of small importance to display a catalogue of writings to be considered as sources of the Geographic History of Vegetables, to briefly add information about those things that are contained by them and pertain to our scope, and to say a word about their merit and worth. Since it may pass beyond the narrow limits of this little work, I hope and promise that I will finish this in another place and time. For this reason, here one may bring into certain classes and divisions only those pertinent writings, to bring forth and advise some things about these single [writings], and to show where we may be able to be taught by many about this subject.

[p. 20] All things that have been composed about matters pertaining to the Geographic History of Vegetables, or comprehend their sources, can be divided in this way:

I. What properly are to be regarded as sources of the Geographic History of Vegetables, and particularly,
   A. Specimens of the Geographic History of Vegetables already published.
   B. Commentaries and sources concerning single heads and parts of the Geographic History of Vegetables.
   C. Sources and writings in which is treated the Geography of Vegetables of single regions of the earth; to which pertain:
      (a) Botanical topographies or Floras.
      (b) Physical topographies.
      (c) Descriptions of travels.
   D. Writings comprising the Geography of single Vegetables (Families, Genera and Species).

II. *Aids to the Geographic History of Vegetables. Here are referred writings that treat Geography, natural History and the History of man.*

[p. 21] *I. Writings that properly are to be regarded as sources of the Geographic History of Vegetables and particularly*

A. *Specimens of the Geographic History of Vegetables already published.*

Authors of this order, whom we hold particularly known thus far, are:

1. Abbé Giraud-Soulavie [1780]

[In 1780] the most celebrated Soulavie proposed the first notion concerning the physical Geography of plants, which he explained more fully in the following book [Giraud-Soulavie 1783].

2. Carolus Ludovicus Willdenow [1797, see Borkhausen 1797]

[p. 22] The natural history of southern France by the most celebrated Soulavie has become known and famous not at all from its merit and its excellence. For although it may contain several paradoxical and almost ridiculous opinions and thoughts, it is exceedingly strong in the philosophical genius with which he contemplates and comprehends all things; which that part of the work cited above that is inscribed *Geographie physique des Plantes* especially openly attests.

Indeed, Menzel [Lesser 1751], Adanson [1763], Forskål [1775] and Zimmermann [1777] already took up certain thoughts on the Geography of Vegetables, which in full justice it is owed to our author to have refined and perfected and it is credited to him as first. In handling this argument the author makes mention of Vegetables particularly of southern France, and inquires into their physico-geographical state excellently and with much erudition. He moreover adds phyto-geographic maps of these regions.

[p. 23] And, even if the author's method of exposition more accurately considered and indeed the order that he has followed in writing this specimen may not always please us, and it is very far from containing the whole complex of things regarding this subject, we cannot but praise and admire the author's exceeding acumen and genius of observation. It is really wonderful how deeply he has followed and penetrated this matter, so that occasionally he connects the weightiest consequences with phenomena of no moment and commonly held to be of little worth.

The other specimen concerning the Geographic History of Vegetables, which has the most celebrated Willdenow as author, is equally and as much worthy of great praise, although the plan of treatment in no way answers to the size and gravity of the object, and also much must be advised against its arrangement, which, as I may say, seems to be merely aphoristic. For the rest I believe that this man was ignorant of the work of Soulavie, or at any rate had no occasion to use it, since many things which he already published are found missing here; in their place, however, he takes notice of much that is new and peculiar. He treats of many migrations of plants, for instance, which the most celebrated Soulavie nearly wholly leaves out.

B. *Commentaries and sources concerning single heads and parts of the Geographical History of Vegetables.*

Writings to be referred here are numerous enough and hardly of small moment, but exceedingly scattered, which, as has already been said above, especially prevails concerning all the sources of this part of natural History [p. 24] and especially concerning this. For it is scarcely possible to set up a rule and regulation that may be unravelled and read from these writings. Very often indeed in writings where the least is expected, we run into notices and observations most grave and of the greatest moment; for instance, in Pennant's *Arctic Zoology* [1784, 1787], Saint-Pierre's *Etudes de la Nature* [1788], Rafn's *Entwurf einer Pflanzenphysiologie* [1798], Girtanner's *Über das Kantische Prinzip für die Naturgeschichte* [1796].

Of writings or commentaries that inquire particularly and separately about single parts of the History of Vegetables, the number is enormous. Among these are especially to be noted [are] de Jussieu [1718], Scheuchzer [1723], Reichel [1750], [p. 25] Linnaeus [1754a, b], Åmann [1756], Flygare [1768], Zinn [1756], Heyne [1785], Forster [1786, 1794], Ramond [1798] [p. 26], Reyiner [1793], Link [1795, 1789], Willdenow [1792, 1797].

C. *Sources and writings in which is treated the Geography of Vegetables of single regions of the earth.*

(a) *Botanical Topographies or Floras.*

Most of these books indeed contain nothing but pure nomenclatures of plants growing in a certain region, arranged according to some system either artificial or, as is said, natural, with the name of the natal place added together. In by far the most, observations of the Phyto-geography of this region are lacking, even though in works of this kind they ought not to be absent. And it must be greatly deplored that the very ones who had the occasion for elucidating, making more certain, and increasing our knowledge in this matter, have done almost nothing of this, but have wholly refrained. Very truly has the most acute St. [p. 27] Pierre expressed himself:

> Il y a des savans, ... nomenclature.

Nevertheless the strongly esteemed Floras of Linneaus [1737–1754], Gmelin [1747–1769], Forskål [1775] [p. 28], Pallas [1784–1788], Forster [1783], Link [1789] and some others are to be excepted.

[p. 29] In Linnaeus's *Bibliotheca botanica* [1747], especially indeed in Banks' *Bibliotheca historico-naturalis* [Dryander 1797], all writings pertaining to this section are reported, digested according to the diverse regions about which they deal. Haller's *Bibliotheca* [1771–1772] also enumerates works of this kind published up to the year 1772, according to chronological order, but dispersed among the remaining botanical writings.

(b) *Physical topographies and*
(c) *Descriptions of travels*

They supply without doubt a perennial spring for elucidating the Geographic History of Vegetables, and especially for augmenting the Phyto-geography of single parts of the Earth.

By no means can one hope that any Topography and description of a Journey pertains to our scope and thus answers to it, as the excellent works of Steller, Crantz, Martens, Dampier, Cook, Mulgrave, Perouse, [p. 30] Bougainville, the Forsters, Labillardiere, Ramond, Saussure, Link, Haenke, Troil, Olaffen, Povelsen, Smith, both Gmelins, Falk, Güldenstaedt, Pallas, Georg, Russel, Thunberg, Levaillant, Sparmann, Patterson, Bruce, Adanson, Isert, Poiret, Volney, Browne, Bartram, Schoepf, and other most celebrated men.

Nevertheless we have set aside no writing of this kind from our hands without any fruit, even though we found little there for our use. For truly, though it must doubtless be lamented and borne with difficulty that very few travellers and topographers have had botanical teaching and erudition, nevertheless they very often supply and restore those things that have been neglected and omitted by those preceding; and I frankly confess that I have dug up more notices in these that afford light to the Geographic History of Vegetables, than I have been able to note in all the so- called botanical works.

D. *Writings comprising the Geography of single Vegetables, Families, Genera and Species.*

Under this heading are comprehended all writings in which something occurs about the present or original abode of a Family, Genus or Species. To this therefore pertain Ichnographies, Describers, and Monographs etc. etc. and indeed, as is apparent, in only that proportion as far as they narrate something about the abodes of Vegetables.

Since the descriptions of plants are of greatest concern to the authors of such books, rarely they indicate, imperfectly and indeterminably, nothing more than [p. 31] the name alone of their natal place.

[p. 32] Among the remaining writings of this kind Georg Foster's commentary on *Arctocarpus*, as an example for constructing a disquisition on this subject, is especially prominent.

II. *Aids to the Geographic History of Vegetables.*

Here are referred writings that treat Geography, Geology, natural History and the History of human Culture.

I believe it is not necessary that I further demonstrate in this place how far writings composed from these doctrines can be brought in as aids to the Geographic History of Vegetables. For the very close connection in which our argument is joined with these is so manifest and open that it does not really need a fuller demonstration.

To special works about those doctrines when enough is already known, it is necessary to add nothing more. One may advise about certain ones concerning common and physical Geography, which most closely touch our argument and especially cohere with it.

In what pertains to geographic notices and indexes that occur in this specimen, I have followed the late Gatterer [1793], so that nevertheless I have inserted and added in their places those things that have been detected or amended from the travels of Perouse [p. 33] Vancouver, Labillardiere, Mungo Park, Browne etc. since the time in which the most celebrated man made the last edition of [his] Geography.

I have made a reckoning of geographic maps with the planiglobes which appeared with Scheider and Weigel at Nuremberg in 1797, and I have also taken into account those things which Pérouse corrected and emended on the Amur coast, Vancouver and Entrecasteaux on the western coast of New Holland [Western Australia], Vancouver and Mal Espina on the coast of New Albion [Pacific coast of North America], Mungo Park, Browne and the most celebrated geographer Rennell in Africa, Georg, Pallas and others in the Russian Empire etc.

In the physical Geography of our planet I have employed the leaders Lulofs [1755], Bergmann [1780] and Lametherie [1795], as much as possible adding those things that have been augmented and explained better in this subject by more recent travellers and investigators of nature.

# References

Adanson, M. (1763). *Familles des plantes*. Paris: Chez Vincent, Imprimeur-Librarie de Mgr le Comte de Provence, rue S. Servin.

Allen, J. A. (1892). The geographical distribution of North American mammals. *Bulletin of the American Museum of Natural History, 4*, 199–243.

Åmann, N. N. (1756). *Flora Alpina*. Uppsala: L.M. Höjer.

Bergmann, T. (1780). *Physikalische Beschreibungen der Erdkugel*. Greifswald: A. F. Röfe.

Borkhausen, M. B. (1797). *Wörterbuch oder Versuch einer Erklärung der vornehmsten Begriffe und Kunstwörter in der Botanik*. Giessen: G.F. Heyer.

Coan, E. V., Kabat, A. R., & Petit, R. E. (2013). *2,400 years of malacology* (10th ed.) American Malacology Society. http://www.malacological.org/publications/2400_malacology.php

de Jussieu, A. (1718). Examen des causes des Impressions des Plants marquées sur certaines pierres des environs de Saint-Chaumont dans le Lionnois. *Histoire de l'Académie Royale des Sciences, 1*, 287–297.

de Lametherie, J- C. (1795). *Thèorie de la Terre*. Paris: Laudan.

de Saint-Pierre, J-H. B. (1797). *Etudes de la Nature. Nouvelle Édition*. Paris: l'Imprimerie-Librarire.

Dryander, J. (1797). *Catalogus Bibliotheca historico-naturalis Josephi Banks*. London: Gul. Bulmer et Soc.

Flygare, J. (1768). *De coloniis plantarum*. Uppsala: L.M. Höjer.

Forskål, P. (1775). *Flora Aegyptiaco – Arabica siue descriptiones plantarum, quas per Aegyptum inferiorem et Arabiam felicem detexit*. Heindeck et Faber, Hauniæ.

Forster, J. R. (1783). *Bemerkungen über Gegenstände der physischen Erdbeschreibung, Naturgeschichte und sittlichen Philosophie auf seiner Reise um die Welt*. Berlin: Haude & Spener.

Forster, G. (1786). *De Plantis esculentis insularum Oceani australis Commentatio botanica*. Berlin: Haude & Spener.

Forster, G. (1794). *Kleine Schriften ein Beytrag zur Völker-und Länderkunde, Naturgeschichte und Philosophie des Leben*s. Berlin: Vossischen Buchhandlung.

Gatterer, J. C. (1793). *Kurzer Begriff der Geographie*. Göttingen: Johann Christian Dieterich.

Giraud-Soulavie, J-L. (1780). La Geographie de la Nature, ou distribution des trois Régnes sur la terre. *Observations sur la Physique, 16*, 63–73.

Giraud-Soulavie, J-L. (1783). *Histoire naturelle de la France meridionale* (Vol. 1). Belin/Paris: Chez J.F. Quillau/Mérigot l'aîné.

Girtanner, C. (1796). *Über das Kantische Prinzip für die Naturgeschichte*. Göttingen: Vandenheok und Ruprecht.

Gmelin, J. G. (1747–1769). *Flora Sibirica sive Historia Plantarum Sibiriae*. Petropoli: Ex Typographia Academiae scientiarum.

Haller, A. (1771–1772). *Bibliotheca botanica, qua scripta ad rem herbariam facientia a rerum initiis recensentur auctore*. Tiguri: Orell, Gessner, Fuessli, et Socc.

Heyne, C. G. (1785). Origines panificii frugumque inventarum initia. *Opuscula Academica collecta et animadversionibus locupletata, 1*, 330–383.

Lesser, F. C. (1751). Nachricht von der Dr. Menzel angegebenen botanischen Geographie. *Physikalische Belustigungen, 1*, 321–327.

Link, H. F. (1789). Florae Gottingensis specimen, sistens vegetabilia saxo calcareo propria. *Usteri's Delectus Opusculorum Botanicorum, 1*, 299–336.

Link, H. F. (1795). Bemerkungen über den Standort der Pflanzen. *Uster's Annalen der Botanik, 14*, 1–17.

Linnaeus, C. (1747). *Bibliotheca botanica*. Halæ Salicae.

Linnaeus, C. (1754a). *Stationes Plantarum*. Defended by Anders Hedenberg. Uppsala: L.M. Höjer.

Linnaeus, C. (1754b). *Coloniae Plantarum*. Defended by Jöns Flygare. Uppsala: L.M. Höjer.

# References

Lulofs, J. (1755). *Einleitung zu der mathematischen und physikalischen Kenntniss der Erdkugel.* Göttingen: A. G. Kästner.

Pallas, P. S. (1784–1788). *Flora Rossica.* Petropoli: J. J. Weitbrecht.

Pennant, T. (1784). *Arctic zoology.* London: Henry Hughs.

Pennant, T. (1787). *Supplement to the Arctic zoology.* London: Henry Hughs.

Rafn, C. G. (1798). *Entwurf einer Pflanzenphysiologie.* Copenhagen: Johann Heinrich Schuboste.

Ramond de Carbonnières, L-F. (1798). Bemerkungen über die Vegetation auf den Gipfeln der höchsten Berge, besonders auf den südlichen Pic der Pyrenäen; wie auch über diejenigen Pflanzen welche mehrere Jahre lang sich unter dem Schnee erhalten können. *Neues polytechnisches Magazin. Eine uswahl aus den wichtigsten französischen Zeitschriften, 1,* 35–53.

Reichel, C. C. (1750). Diatribe de Vegetabilibus petrefactis. Wittenberg: Zimmermann.

Reynier, L-F. (1793). De l'influence du Climat sur la forme et la nature des Végétaux. *Observations sur la Physique, 43,* 399–420.

Scheuchzer, J. J. (1723). *Herbarium diluvianum.* Lugduni Batavorum: Sumptibus Petri Vander Aa.

Willdenow, C. L. (1792). *Grundriss der Kräuterkunde.* Berlin: Haude and Spener.

Willdenow, C. L. (1797). Beyträge zur geographischen Geschichte des Pflanzenreichs. *Uster's Annalen der Botanik, 22,* 1–13.

Willdenow, C. L. (1810). *Grundriss der Kräuterkunde.* Berlin: Haude and Spener.

Willdenow, C. L. (1811). *The principles of botany and of vegetable physiology.* London: William Blackwood.

Zimmermann, E. A. W. (1777). *Specimen zoologiae geographicae, Quadrupedum domicilia et migrationes sistens.* Leiden: Theodorum Haak.

Zimmermann, E. A. W. (1778–1783). *Geographische geschichte des menschen, und der allgemein verbreiteten vierfüssigen thiere.* Leipzig: Weygandschen Buchhandlung.

Zinn, J. G. (1756). Von dem Ursprunge der Pflanzen. *Hamburgisches Magazin, oder gesammlete Schriften, zum Unterricht und Vergnügen, 16,* 339–355.

The manufacturer's authorised representative in the EU is Springer Nature Customer Service Centre GmbH, Europaplatz 3, 69115 Heidelberg, Germany. If you have any concerns regarding our products, please contact ProductSafety@springernature.com

Printed and bound by CPI Group (UK) Ltd, Croydon, CR0 4YY

25/03/2026

02078197-0015